Contents

Nelson Thornes and AQA

Nelson Thornes has worked in partnership with AQA to ensure that this book and the accompanying online resources offer you the best support for your GCSE course.

All AQA endorsed resources undergo a thorough quality assurance process to ensure that their contents closely match the AQA specification. You can be confident that the content of materials branded with AQA's 'Exclusively Endorsed' logo have been written, checked and approved by AQA senior examiners, in order to achieve AQA's exclusive endorsement.

The print and online resources together unlock blended learning; this means that the links between the activities in the book and the activities online blend together to maximise your understanding of a topic and help you achieve your potential.

These online resources are available on kerboodle! which can be accessed via the internet at www.kerboodle.com/live, anytime, anywhere.

If your school or college subscribes to kerboodle! you will be provided with your own personal login details. Once logged in, access your course and locate the required activity.

For more information and help on how to use kerboodle! visit www.kerboodle.com.

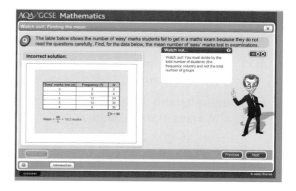

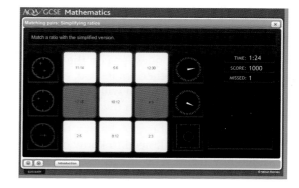

How to use this book

To help you unlock blended learning, we have referenced the activities in this book that have additional online coverage in *kerboodle!* by using this icon:

The icons in this book show you the online resources available from the start of the new specification and will always be relevant.

In addition, to keep the blend up-to-date and engaging, we review customer feedback and may add new content onto *kerboodle!* after publication.

Welcome to GCSE Mathematics

This book has been written by teachers and examiners who not only want you to get the best grade you can in your GCSE exam, but also to enjoy maths. It covers all the material you will need to know for AQA GCSE Mathematics Unit 1 Higher. This unit allows you to use a calculator, so you will be able to use this most of the time throughout this book. Look out for calculator or non-calculator symbols (shown below) which may tell you whether to use a calculator or not.

In the exam, you will be tested on the Assessment Objectives (AOs) below. Ask your teacher if you need help to understand what these mean.

AO1 recall and use your knowledge of the prescribed content

AO2 select and apply mathematical methods in a range of contexts

AO3 interpret and analyse problems and generate strategies to solve them.

Each chapter is made up of the following features:

Objectives

The objectives at the start of the chapter give you an idea of what you need to do to get each grade. Remember that the examiners expect you to do well at the lower grade questions on the exam paper in order to get the higher grades. So, even if you are aiming for a Grade A you will still need to do well on the Grade D questions on the exam paper.

On the first page of every chapter, there are also words that you will need to know or understand, called Key Terms. The box called 'You should already know' describes the maths that you will have learned before studying this chapter. There is also an interesting fact at the beginning of each chapter which tells you about maths in real life.

Learn...

The Learn sections give you the key information and examples to show how to do each topic. There are several Learn sections in each chapter.

Practise...

Questions that allow you to practise what you have just learned.

D The bars that run alongside questions in the exercises show you what grade the question is aimed at. This will give you an idea of what grade you're working at. Don't forget, even if you are aiming at a Grade A, you will still need to do well on the Grades D–B questions.

These questions are Functional Maths type questions, which show how maths can be used in real life.

? These questions are problem solving questions, which will require you to think carefully about how best to answer.

These questions are harder questions.

These questions should be attempted **with** a calculator.

These questions should be attempted **without** using a calculator.

Assess

End of chapter questions written by examiners. Some chapters feature additional questions taken from real past papers to further your understanding.

Hint

These are tips for you to remember whilst learning the maths or answering questions.

AQA Examiner's tip

These are tips from the people who will mark your exams, giving you advice on things to remember and watch out for.

Bump up your grade

These are tips from the people who will mark your exams, giving you help on how to boost your grade, especially aimed at getting a Grade C.

Consolidation

The consolidation chapter allows you to practise what you have learned in previous chapters. The questions in these chapters can cover any of the topics you have already seen.

1 Fractions and decimals

Objectives

Examiners would normally expect students who get these grades to be able to:

D

find one quantity as a fraction of another

solve problems involving fractions

C

add and subtract mixed numbers

multiply and divide fractions

round numbers to significant figures

B

find upper and lower bounds

A/A*

use upper and lower bounds in calculations.

Did you know?

Fractions and decimals in music

Why do musical instruments sound different from each other? A lot of it has to do with fractions. Pluck a guitar string and you hear a certain note, but mixed in with the main note there are quieter, higher notes. You can bring out one of these notes, called 'harmonics', by lightly touching the string halfway along when you pluck it. In fact a vibrating string naturally produces harmonics that correspond to $\frac{1}{2}$, $\frac{1}{3}$, $\frac{1}{4}$, $\frac{1}{5}$ ... of its length all at the same time. The tubes of air in wind instruments, like trumpets and saxophones, do the same. The construction of an instrument, and the way it's played, affect how loud the different harmonics are, compared with one another. That's why a violin sounds different from a clarinet.

Decimals come into music too. The pitch of a note (how high or low it is) depends on the number of times per second it makes the air (and therefore your eardrums) vibrate. The faster the vibrations, the higher the note. Play a note on a keyboard, and then the nearest note above it: the higher note causes vibrations approximately 1.05946... times faster than the lower note.

Key terms

numerator
denominator
equivalent fraction
mixed number
rounding
significant figures
lower bound
upper bound

You should already know:

✔ what equivalent fractions are

✔ how to add and subtract simple fractions

✔ how to calculate fractions of quantities

✔ how to express simple decimals and percentages as fractions

✔ how to order decimals.

Reminder: You will be allowed to use a calculator in this unit. Make sure you know how to use the fraction (or a^b_c) button on your calculator.

Calculators will simplify your fractions as well. Just put in the fraction using the fraction button then press = to get the simplest form.

Learn... 1.1 One quantity as a fraction of another

To work out one quantity as a fraction of another, write the first quantity as the **numerator** and the second as the **denominator** then simplify the fraction.

To work out 35 as a fraction of 50, write 35 out of 50 as a fraction, $\frac{35}{50}$, then simplify to $\frac{7}{10}$

So 35 is seven-tenths of 50.

$\frac{35}{100}$ and $\frac{7}{10}$ are equivalent fractions.

Example: What is 25 cm as a fraction of 2 m?

Solution: 2 m is 200 cm.

So the fraction is $\frac{25}{200}$,

which simplifies to $\frac{1}{8}$.

Practise... 1.1 One quantity as a fraction of another

D C B A A*

1 What is:

a   15p as a fraction of 30p

b   15p as a fraction of £3

c   20 minutes as a fraction of an hour

d   20 minutes as a fraction of two hours

e   100 g as a fraction of 1.3 kg

f   200 g as a fraction of 1.3 kg?

2 Kevin says '50p as a fraction of £500 is $\frac{1}{10}$.' Is Kevin right? Explain your answer.

3 Here is a list of the heights of 12 students, measured to the nearest centimetre and arranged in order.

155 cm   159 cm   161 cm   162 cm   162 cm   165 cm
167 cm   169 cm   172 cm   174 cm   175 cm   177 cm

a   What fraction of students have heights:

i   less than 166 cm

ii   more than 166 cm

iii   between 161.5 cm and 173 cm?

b   What fraction of students have heights below 161.5 cm?

4 a   The students in a class of 28 take a test.
What fraction of them passed the test if the number passing was as follows:

i   14   ii   12   iii   13   iv   18   v   20?

b   If the fraction passing the test was $\frac{3}{4}$, how many students passed?

c   Explain why you should never get an improper (top-heavy) fraction in a question like this.

5 In a dance class there are 25 women and 15 men.

a   What is the fraction of men in the class?

b   What is the fraction of women in the class?

D

6   Here is a list of test marks of a class of 30 students, arranged in order.

a   What fraction of the students got under 40 marks?

b   What fraction of the students got a mark between 60 and 70?

c   The pass mark was 50 marks. What fraction of the students passed the test?

d   What should the pass mark be for two-thirds of the students to pass the test?

e   What mark separates the top tenth of the class from the rest?

```
22   25
30   33   37
42   43   46   46
53   54   55   55   56
61   61   63   64   64   67   68   68   69
73   75   78   79
81   87
95
```

7   What is:

a x as a fraction of $10x$

b $5b$ as a fraction of $25b$

c x as a fraction of x^2?

8

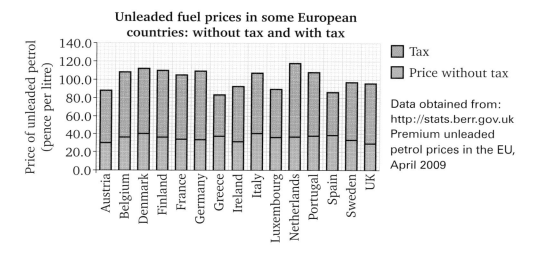

Unleaded fuel prices in some European countries: without tax and with tax

Data obtained from:
http://stats.berr.gov.uk
Premium unleaded
petrol prices in the EU,
April 2009

The graph shows how the total price paid for unleaded petrol is made up.

a   Estimate the fraction of the total price paid in tax in the UK.

b   Which country pays the highest fraction in tax?

c   Which country pays the lowest?

d   Would it affect your answers if the prices were in euros instead of in pence? Explain your answer.

9   Here are 32 students' test marks. Three-quarters of the students passed the test. What could the pass mark have been?

```
14
23   29
33   34   36
46   47
54   53   56   56   56   59   59
60   60   61   64   65   67   68   68
70   71   71   73   75   76   77
80   82
```

Learn...   **1.2 Calculating with fractions**

In this unit you will be using a calculator to work with fractions. There is more work on fraction calculations in Unit 2.

With a calculator, you can use the fraction button

▭ or $a\frac{b}{c}$

to do all fraction calculations.

For example, to do the calculation $\frac{3}{4} + \frac{2}{5}$, do this:

▭ 3 ▸ 4 ▸ ➕ ▭ 2 ▸ 5 🟰

or   3 $a\frac{b}{c}$ 4 ➕ 2 $a\frac{b}{c}$ 5 🟰

which gives the answer $\frac{23}{20}$ or $1\frac{3}{20}$

> ### AQA *Examiner's tip*
> Make sure you know how to put **mixed numbers** into your calculator using the fraction button and how to convert between mixed numbers and top-heavy fractions.

Example: Dave makes fleece hats to sell on his market stall. Each hat needs $\frac{3}{8}$ of a yard of fabric. How many hats can he make from $2\frac{1}{2}$ yards of fabric?

Solution: To find how many hats can be made, divide $2\frac{1}{2}$ by $\frac{3}{8}$

▭ 5 ▸ 2 ▸ ➗ ▭ 3 ▸ 8 🟰 $\frac{20}{3}$ 🟰 $6\frac{2}{3}$

So Dave can make 6 hats with enough fabric for $\frac{2}{3}$ of a hat left over.

Practise...   **1.2 Calculating with fractions**   

D

1   Zeb spends $\frac{2}{5}$ of his pocket money on clothes, $\frac{1}{3}$ on CDs and $\frac{1}{4}$ on going to the cinema. What fraction of his pocket money has he spent altogether?

2   A group of students and teachers are on a trip.

$\frac{1}{4}$ of the group are Year 9. Of the remainder, $\frac{2}{5}$ are Year 10.
There are also 19 students from Year 11 and 8 teachers.
How many people are in the group altogether?

3   There are red, blue and green discs in a bag.

$\frac{1}{3}$ of the discs are red.

$\frac{2}{5}$ of the discs are blue.

There are 12 green discs.

How many red discs are there?

⚠ 4   Here is a fraction pattern.

$\frac{1}{2}$ $\qquad\qquad\qquad = \frac{1}{2}$

$\frac{1}{2} + \frac{1}{4}$ $\qquad\qquad = \frac{3}{4}$

$\frac{1}{2} + \frac{1}{4} + \frac{1}{8}$ $\qquad = \frac{7}{8}$

$\frac{1}{2} + \frac{1}{4} + \frac{1}{8} + \frac{1}{16}$ $\quad =$

$\frac{1}{2} + \frac{1}{4} + \frac{1}{8} + \frac{1}{16} + \frac{1}{32} =$

a   Find the sums for the last two rows of the pattern.

b   What is the sum of the tenth row of the pattern?

c   What do you think will happen to the sum of the rows as you keep adding more and more terms?

5 The same T-shirt is available in two different shops.

Which shop gives the better price? Show how you worked out the answer.

6 In a sale, everything is offered with 'one quarter off'. Work out the sale prices of these items, rounding to the nearest penny if necessary.

a A sweater costing £24

b A dress costing £69.99

c A pair of trousers costing £24.99

d A pen costing 89p

7 In America, dress material is sold in yards and fractions of yards. A jacket needs $2\frac{1}{2}$ yards of material and a skirt needs $1\frac{3}{4}$ yards.

a How much material is needed altogether?

b The material costs $15 a yard. How much will the material for the jacket and skirt cost?

8 A video game runs at a frame rate of 35 frames a second. The movement of a car takes 15 frames ($\frac{15}{35}$ of a second). Going round a corner and stopping takes a further 30 frames. How much time is taken by these two scenes?

9 A recipe for biscuits needs two-thirds of a cup of sugar.
A pudding needs $1\frac{3}{8}$ cups of sugar.

Lucy has two cups of sugar.

Does she have enough to make the biscuits and the pudding?

10 Thirty years ago in the UK, lengths used to be measured in yards and fractions of a yard. Here is a problem from that time.

A room measures $4\frac{1}{4}$ yards by $3\frac{3}{8}$ yards. The door is $\frac{7}{8}$ of a yard wide.
What length of skirting board is needed for this room?

11 *A* and *B* are two whole numbers.

a One-fifth of *A* is equal to one-quarter of *B*.

i Which number is bigger, *A* or *B*? Explain how you know.

ii Find possible values for *A* and *B*. How many pairs are there?

iii For each pair, work out the fraction $\frac{A}{B}$. What do you notice?

b Three-quarters of *A* is equal to two-thirds of *B*. Work out the fraction $\frac{A}{B}$

12

Here are two strips of card.

The top one is divided into quarters. The bottom one is divided into sixths.

Work out the total width of the diagram.

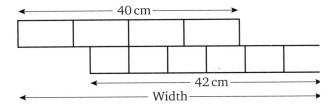

Learn... 1.3 Rounding

You have already seen how to round numbers to decimal places. For example,

1.2832 correct to one decimal place is 1.3

1.2832 correct to two decimal places is 1.28

and to three decimal places is 1.283

You also need to be able to round them to a certain number of **significant figures**.

12 832 has five significant figures.

Rounding it to three significant figures (s.f.) gives 12 800.

Rounding it to 2 s.f. gives 13 000 because 12 832 is closer to 13 000 than it is to 12 000.

(In each case the zeros are not significant figures because they only show the place values of the other digits.)

Note that zeros *can* sometimes be significant figures.

Rounding 1.2832 to 2 s.f. gives 1.3

> AQA **Examiner's tip**
>
> Make sure you know the difference between significant figures and decimal places.
> 1.3 has two significant figures but only one decimal place.

Example: Round these numbers to **a** one significant figure, **b** two significant figures, **c** three significant figures.

i	16 784	iv	1500	vii	15.3142
ii	20 259	v	1.6784	viii	0.1500
iii	153 142	vi	2.0259	ix	0.00005748

Solution:

a
i	20 000	iv	2000	vii	20
ii	20 000	v	2	viii	0.2
iii	200 000	vi	2	ix	0.00006

b
i	17 000	iv	1500	vii	15
ii	20 000	v	1.7	viii	0.15
iii	150 000	vi	2.0	ix	0.000057

c
i	16 800	iv	1500	vii	15.3
ii	20 300	v	1.68	viii	0.150
iii	153 000	vi	2.03	ix	0.0000575

Practise... 1.3 Rounding

D C B A A*

1 **a** Round these numbers to one significant figure.

i	1725.251	**iv**	5.5921
ii	525.48	**v**	153.29
iii	25.500	**vi**	0.0135

b Round these numbers to two significant figures.

i	1725.251	**iv**	5.5921
ii	525.48	**v**	153.29
iii	25.500	**vi**	0.0135

> **Bump up your grade**
>
> To get a Grade C you need to know how to round to a given number of significant figures.

2 Here are some mistakes made with rounding.
Write a correct statement for each one.

a 734 522 to three significant figures is 735

b 1442 to the nearest 10 is 1450

c 25.60 to the nearest whole number is 26.00

3 John and Kieran count their collections of marbles.

John has 18 more marbles than Kieran.

Rounded to the nearest 10, John has 50 marbles and Kieran has 40 marbles.

Write down the possible numbers of marbles each boy might have.

4 Anne buys 53.4 litres of petrol at 99 pence per litre.
How much will she have to pay?

5 Mita buys 23.2 litres of diesel and pays £23.90.

a What is the price per litre of the diesel?

b Explain why your answer is probably not exact.

6 Judy's weight is 56.67 kg.

a What is her weight to the nearest:

i kilogram

ii half kilogram

iii 100 grams?

b Body mass index (BMI) is worked out by dividing weight (in kilograms) by the square of height (in metres).

That is, $\text{BMI} = \dfrac{\text{weight}}{(\text{height})^2}$

Judy is 1.47 metres tall.

i Calculate Judy's body mass index.

A person with a BMI of over 25 is considered to be overweight.

ii How much weight, to the nearest kilogram, should Judy lose?

Learn... 1.4 Upper and lower bounds

Continuous measurements can never be completely accurate and it is helpful to know how accurate they are.

Suppose that the weight of a person is measured as 62 kg to the nearest kilogram on bathroom scales.

This means that the weight is nearer to 62 kg than it is to either 61 kg or 63 kg. It can be half a kilogram either side of 62 kg.

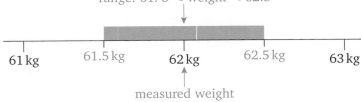

The real weight can be anywhere in the range: $61.5 \leqslant$ weight < 62.5

61 kg   61.5 kg   62 kg   62.5 kg   63 kg

measured weight

If W kg is the weight, $61.5 \leqslant W < 62.5$

61.5 kg is the **lower bound** of the weight and 62.5 kg is the **upper bound.**

A weight less than 61.5 kg would be rounded to 61 kg.

A weight greater than 62.5 kg would be rounded to 63 kg.

Example: The weight of a package, P kg, is 4 kg to the nearest kilogram.

    a What are the upper and lower bounds of the weight?

    b Copy and complete this statement: ___ $\leqslant P <$ ___

Solution: **a** 4 kg to the nearest kilogram means it can be half a kilogram either way.
So the upper bound is 4.5 kg and the lower bound is 3.5 kg.

    b $3.5 \leqslant P < 4.5$

Practise... 1.4 Upper and lower bounds   c

1 These numbers have been rounded to the nearest integer.
Write down their upper and lower bounds.

    a 15   **b** 157   **c** 100   **d** 4   **e** 0

2 Write down the upper and lower bound of these volumes measured to the nearest cubic centimetre (cm^3).

    a $100 \, cm^3$ **b** $15 \, cm^3$ **c** $245 \, cm^3$ **d** $1000 \, cm^3$ **e** $500 \, cm^3$

3 The distance between two towns is 255 km correct to the nearest 5 km.

What is the upper bound of the distance?

4 Anne's weight is 66 kg correct to the nearest half kilogram.

What is the least Anne could possibly weigh?

5 These lengths are correct to the nearest half centimetre.

What are the upper and lower bounds of each length?

    i 10.5 cm   **ii** 7.0 cm   **iii** 23.5 cm   **iv** 15.0 cm

6 A breakfast cereal packet says that the size of a serving of cereal is 37.5 g.

Assuming that this is measured correct to the nearest half gram, what are the upper and lower bounds of the weight of a serving of this cereal?

7 An adult's daily recommended intake of salt is 6 g.

Is an adult who takes 6.5 g of salt within the recommendation?

What assumption do you have to make to answer this question?

Learn... 1.5 Calculating with bounds

It is important to consider possible inaccuracies in measurements when combining measurements. For example, a packet could contain 20 chocolate bars, each weighing 35 g to the nearest gram.

If all the bars weighed the least they possibly could, the lower bound of the total weight in the pack of 20 would be $20 \times 34.5 \, g = 690 \, g$

If all the bars weighed the most they possibly could, the upper bound of the total weight in the pack of 20 would be $20 \times 35.5 \, g = 710 \, g$ which is quite a difference!

Example: Cartons contain 75 ml of fruit juice correct to the nearest half ml.

What are the maximum and minimum amounts of juice in a pack of 12 cartons of juice?

Solution: The least possible volume of juice in a carton is 74.75 ml and the maximum is 75.25 ml.

The minimum volume of juice in the pack is $12 \times 74.75 \, ml = 897 \, ml$

The maximum volume of juice in the pack is $12 \times 75.25 \, ml = 903 \, ml$

Example: A small bolt for a computer must be 5 mm long with a tolerance of 10%.
What is the maximum possible length of the bolt?

Solution: A tolerance of 10% means that the bolt can be up to 10% less than 5 mm and up to 10% more than 5 mm.

So the maximum possible length is 5 mm + 10% of 5 mm = 5 mm + 0.5 mm = 5.5 mm

---

In calculating the maximum or minimum value of a quantity you have to think carefully about which bound to use: the upper bound or the lower bound.

If you need the **maximum** value:

maximum of $(a + b)$ = maximum of a + maximum of b

maximum of $(a - b)$ = maximum of a − minimum of b

maximum of $a \times b$ = maximum of a × maximum of b

maximum of $\dfrac{a}{b} = \dfrac{\text{maximum of } a}{\text{minimum of } b}$

If you need the **minimum** value:

minimum of $(a + b)$ = minimum of a + minimum of b

minimum of $(a - b)$ = minimum of a − maximum of b

minimum of $a \times b$ = minimum of a × minimum of b

minimum of $\dfrac{a}{b} = \dfrac{\text{minimum of } a}{\text{maximum of } b}$

Example: Find the minimum number (the number you can be sure of getting) of short pieces of string of length 33 cm that can be cut from a long piece of length 453 cm. Lengths are rounded to the nearest centimetre.

Solution: You need to use the minimum possible length of the long string and the maximum possible length of the short pieces (the 'worst case scenario') in the calculation.

$452.5 \div 33.5 = 12.7$, so the minimum number of short pieces that can be cut is 12.

> **AQA Examiner's tip**
>
> Normally 12.7 rounded to the nearest integer would be 13, but here you have to ignore any amount less than 1 to work out the number of pieces with the required length. (You cannot make 12.7 pieces into 13 whole pieces because the 13th piece would be too short.)

Practise...   1.5 Calculating with bounds   

B

A

A*

1   A rectangle has length 10.5 cm and width 7.0 cm, both correct to the nearest half centimetre.
What is the maximum length of the perimeter of the rectangle?

2   Alfie says 'If a length is 93 cm to the nearest cm, the maximum possible length is 93.49 cm.'

Is Alfie correct? Explain your answer.

3   An athlete runs 100 m (measured to the nearest 5 cm) in a time of 13.1 seconds (measured to the nearest tenth of a second).

What is the maximum value of the athlete's average speed for this run?

4   Keith drives a distance of 120 miles, correct to the nearest 10 miles, at an average speed of 45 miles per hour, correct to the nearest 5 miles per hour.

What is the maximum length of time the journey will take?

5   A packet of breakfast cereal contains 450 grams of cereal, correct to the nearest 5 grams. What is:

a   the maximum   **b**   the minimum

number of 35 g servings, correct to the nearest 5 grams, that you can serve from this packet?

6   15 cartons each weighing 12.5 kg correct to the nearest 0.5 kg are to be lifted by a hoist that can take a maximum load of 190 kg.

Is it safe to use this hoist to lift these cartons?

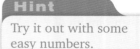

7   Choco bars weigh 53 g to the nearest gram.
Toffo bars weigh 72 g to the nearest gram.

Calculate the maximum and minimum differences between the weights of the two types of bars.

! 8 a, b, c and d are four positive numbers, each rounded to the nearest whole number. $a > b$ and $c > d$.

Mina has to find the maximum possible value of the answer to the calculation

$$\frac{a - b}{c - d}$$

Which of the numbers a, b, c and d should she give their maximum possible value and which should she give their minimum possible value?

> **Hint**
> Try it out with some easy numbers.

 9   A ball bearing has a diameter of 5.3 mm with a 5% tolerance. (It could be 5% more or 5% less.) The ball bearing's diameter must be correct to the nearest 0.5 mm. Does this ball bearing fit the specification?

 10   A cylindrical bar has to fit into a circular tube. The diameter of the bar is 5.3 mm correct to the nearest tenth of a millimetre and the diameter of the tube is 5.4 mm with a 10% tolerance.

Is it certain that the bar will fit into the tube?

 11   Kathryn's weight is 53.8 kg to the nearest 100 g. 1 kg = 2.20 pounds correct to three significant figures. What is the upper bound of Kathryn's weight in pounds?

Assess

1   A train arrived at a station 30 minutes late.
What fraction is this of the journey time of $2\frac{1}{2}$ hours?

D

2   Catriona buys a car costing £3500.

There is a reduction of one-fifth of the price.
She pays three-eighths of the reduced price as deposit.

How much is the deposit?

3   The table shows the mean temperature in degrees Celsius each month in
Anchorage, Alaska, USA.

Jan	Feb	Mar	Apr	May	Jun	Jul	Aug	Sep	Oct	Nov	Dec
−10	−7	−4	2	8	12	15	14	9	1	−6	−9

a   For what fraction of the months is the mean temperature above zero?

b   For what fraction of the months is the mean temperature more than 5 degrees below
zero?

4   The bar chart shows the percentage of national income spent on education by
EU countries in 2002.

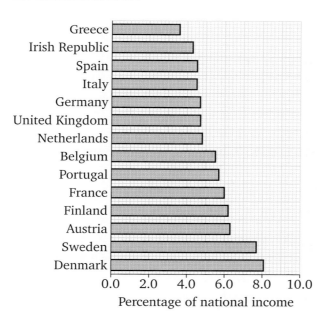

Source: Education at a Glance 2002,
Organisation for Economic
Co-operation and Development

a   What fraction of countries spent less than 5% of their national income on education?

b   What fraction of countries spent more on education than the United Kingdom?

5   Work out:

a   one-third of four-fifths of £30

b $2\frac{2}{9} + 4\frac{1}{2}$

c   the number that, when reduced by one-third, becomes 14

d   the price that, when increased by one-quarter, becomes £26.50.

C

6 Write down three numbers that:
    a   to one significant figure round to   **i** 100   **ii** 0.01
    b   to two significant figures round to   **i** 120   **ii** 0.12

7 An amount of money is £49 to the nearest pound.
What is the greatest amount it could be? What is the least?

B

8 If a girl's age is correctly stated as 12, what is the maximum age she could be?

9 A length l cm is measured as 15 cm correct to the nearest half centimetre.
Copy and complete the statement ___ $\leqslant l <$ ___

10 Rohan puts petrol in his car. Petrol is £0.98 a litre and he pays £49.90 to fill the tank.
How many litres does he buy?
How accurate is your answer?

A

11 Lyra's height is 1.54 m and Louisa's is 1.61 m. Both heights are measured to the nearest centimetre.
What is the maximum possible difference between Lyra's height and Louisa's height?

12 Bags of sweets are supposed to weigh 100 g correct to the nearest 5 grams.
They are filled with sweets that weigh 8 g correct to the nearest gram.
If 13 sweets are put into each bag can you be sure the weight is enough?

13 The maximum weight of people that can safely travel in a lift is exactly 350 kilograms.
Four people are already in the lift. Their weights to the nearest kilogram are 63 kg, 65 kg, 72 kg and 75 kg.
    a   What is the maximum total weight of the people in the lift?
Rob wants to join them in the lift. Rob weighs 70 kg to the nearest 10 kilograms.
    b   Can Rob safely join them in the lift? Give reasons for your answer.

A*

14 A factory makes ball bearings with a diameter of 4 mm correct to the nearest millimetre. They are spheres of stainless steel. The volume of stainless steel available is 525 cubic centimetres (cm^3) to the nearest 5 cm^3.
What is the minimum number of ball bearings that can be made?
(The volume, V, of a sphere of radius r is given by $V = \frac{4}{3}\pi r^3$)

15 An athlete completes a mile (1760 yards, measured to the nearest tenth of a yard) in a race in 4 minutes, measured to the nearest tenth of a second. A yard is 0.914 m to the nearest millimetre.
What is the maximum possible value of the speed of the athlete in metres per second?

AQA Examination-style questions

1 A book has a front and back cover and 100 pages.
The front and back cover are each 0.8 millimetres thick when measured to one decimal place.
Each page is 0.15 millimetres thick when measured to two decimal places.

Calculate the minimum thickness of the book.
You **must** show your working.

(2 marks)

AQA 2007

Objectives

Examiners would normally expect students who get these grades to be able to:

C

use index notation and index laws for positive and negative powers including $10^3 \times 10^5$ and $\dfrac{10^3}{10^7}$

B

convert between ordinary and standard index form numbers

use standard index form for calculations.

Did you know?

A googol

A googol is the number 10^{100}, i.e. the digit 1, followed by one hundred zeros. It is such a large number that it is bigger than the number of atoms in the known universe! The term googol was introduced by nine-year-old Milton Sirotta who was the nephew of American mathematician Edward Kasner. He was asked to think of a name for the very large number 10^{100}, which was written about in the book *Mathematics and the Imagination*.

Maybe you can invent a name for a special number.

You should already know:

✔ how to multiply numbers

✔ how to calculate squares and square roots

✔ how to calculate cubes and cube roots

✔ how to use function keys on a calculator for powers

✔ how to use algebra.

Key terms

indices
index
power
exponent
standard index form

Learn... 2.1 Rules of indices

Index (or power)

10^3

In words say '10 to the power 3'.

Base

The **index** (or **power** or **exponent**) tells you how many times the base number is to be multiplied by itself.

This means that 10^3 tells you that 10 (the base number) is to be multiplied by itself 3 times (3 here is the index or power).

So $10^3 = 10 \times 10 \times 10 = 1000$

Hint

You can use the $x^\blacksquare$ button on your calculator.

[1] [0] [$x^\blacksquare$] [3] [=] 125

Rules of indices

In general:
$$10^m \times 10^n = 10^{m+n}$$
$$10^m \div 10^n = 10^{m-n}$$
$$(10^m)^n = 10^{mn}$$
$$10^{-m} = \frac{1}{10^m}$$
$$10^0 = 1$$

Bump up your grade

You will need to use index notation and index laws for positive and negative powers for an award of Grade C.

Example: Simplify:

a $10^3 \times 10^2$

b $\dfrac{10^2}{10^5}$

c $(10^7)^2$

Solution:

a $10^3 \times 10^2$
$= 10^{(3+2)}$
$= 10^5$

b $\dfrac{10^2}{10^5}$
$= 10^2 \div 10^5$
$= 10^{(2-5)}$
$= 10^{-3}$

c $(10^7)^2$
$= 10^{7 \times 2}$
$= 10^{14}$

Practise... 2.1 Rules of indices   D C B

1 Calculate:

a 10^2

b 10^6

c 10^0

d $10^2 + 10^2$

e $10^3 - \sqrt{100}$

f $\sqrt[3]{1000}$

2 Write in index notation:

a $10 \times 10 \times 10 \times 10$

b $10 \times 10 \times 10 \times 10 \times 10 \times 10 \times 10$

c 10×10

d 10

e $\dfrac{1}{10}$

f $\dfrac{1}{1000}$

3 Are the following statements true or false? Give a reason for your answer.

a $10^2 = 20$

b $10^{-1} = -10$

c $10^{50} \times 10^{50} = 10^{100}$

d $1\,000\,000^0 = 0$

4 Simplify the following, leaving your answer in index form.

a $\dfrac{10^7 \times 10^3}{10^6}$

b $\dfrac{10^7 \times 10^4 \times 10^3}{10^6}$

5 Simplify the following, leaving your answer in index form. Make sure your answer uses a positive power of 10.

a $10^6 \times 10^5$ **f** $(10^2)^5$ **k** $(10^{-4})^{-2}$

b $10^8 \times 10^3$ **g** $(10^3)^2$ **l** $10^5 \div 10^8$

c $\dfrac{10^7}{10^3}$ **h** $(10^2)^3$ **m** $(10^4)^2 \times (10^3)^{-2}$

d $10^{11} \div 10^5$ **i** $(10^{-3})^2$ **n** $10^2(10^3 + 10^{-3})$

e $10^7 \div 10^{10}$ **j** $(10^{-2})^3$ **o** $10^{-5} \times 10^{-2} \div 10^{-7}$

 Learn... 2.2 Standard index form

Standard index form is a shorthand way of writing very large and very small numbers.

Standard form numbers are always written as follows.

a power of 10

$$A \times 10^n$$

a number between 1 and 10

Converting from standard index form

To convert from standard index form to ordinary form, use the following information.

$10^1 = 10$

$10^2 = 10 \times 10 = 100$

$10^3 = 10 \times 10 \times 10 = 1000$

$10^4 = 10 \times 10 \times 10 \times 10 = 10\,000$

$10^5 = 10 \times 10 \times 10 \times 10 \times 10 = 100\,000$

$10^6 = 10 \times 10 \times 10 \times 10 \times 10 \times 10 = 1\,000\,000$ (1 million)

$10^{-1} = \dfrac{1}{10^1} = \dfrac{1}{10} = 0.1$

$10^{-2} = \dfrac{1}{10^2} = \dfrac{1}{100} = 0.01$

$10^{-3} = \dfrac{1}{10^3} = \dfrac{1}{1000} = 0.001$

$10^{-4} = \dfrac{1}{10^4} = \dfrac{1}{10\,000} = 0.0001$

$10^{-5} = \dfrac{1}{10^5} = \dfrac{1}{100\,000} = 0.00001$

$10^{-6} = \dfrac{1}{10^6} = \dfrac{1}{1\,000\,000} = 0.000001$

> **AQA Examiner's tip**
>
> Remember that 10^{-1} is the same as dividing by 10, and 10^{-2} is the same as dividing by 10^2 etc.

Converting to standard index form

To convert to standard index form from ordinary form:

Write your number in the form $A \times 10^n$ where A is a number between 1 and 10.

Example: Convert these ordinary form numbers into standard index form.

    a 65 000   **b** 0.000000572   **c** 4.2

Solution: **a** 65 000

$A = 6.5$

so $65\,000 = 6.5 \times 10\,000 = 6.5 \times 10^4$

b 0.000000572

$A = 5.72$

so $0.000000572 = 5.72 \times 0.0000001$
$$= 5.72 \times 10^{-7}$$

c 4.2

$A = 4.2$, so $4.2 = 4.2 \times 1 = 4.2 \times 10^0$

AQA Examiner's tip

It is helpful to count the number of places the decimal point needs to move to form a number between 1 and 10.
$0.\overset{\frown}{0}\overset{\frown}{0}\overset{\frown}{0}\overset{\frown}{0}\overset{\frown}{0}\overset{\frown}{0}\overset{\frown}{5}72 = 5.72 \times 10^{-7}$

Example: **a** Convert these standard index form numbers into ordinary form.

    i 2×10^2

    ii 6.82×10^5

    iii 3.001×10^3

b Convert these standard index form numbers into ordinary form.

    i 2×10^{-2}

    ii 6.82×10^{-5}

    iii 3.001×10^{-3}

Solution: **a** **i** $2 \times 10^2 = 2 \times 100$
$$= 200$$

    ii $6.82 \times 10^5 = 6.82 \times 100\,000$
$$= 682\,000$$

    iii $3.001 \times 10^3 = 3.001 \times 1000$
$$= 3001$$

b **i** $2 \times 10^{-2} = 2 \times 0.01$
$$= 0.02$$

    ii $6.82 \times 10^{-5} = 6.82 \times 0.00001$
$$= 0.0000682$$

    iii $3.001 \times 10^{-3} = 3.001 \times 0.001$
$$= 0.003001$$

AQA Examiner's tip

Make sure you can add, subtract, multiply and divide standard index form numbers using your calculator.

Practise... 2.2 Standard index form

B

1 Write the following ordinary numbers in standard index form.

    a 3700   **e** 35   **i** 0.00000000002

    b 23 000 000   **f** 0.005   **j** 0.5

    c 200 200   **g** 0.13

    d 8 500 000 000   **h** 0.000000178

2   Write the following standard index form numbers as ordinary numbers.

a 7×10^3 e 7.6635×10^1 i 3.086×10^{-4}

b 7.0×10^3 f 5.1×10^8 j 6.6×10^{-10}

c 4.2×10^4 g 3×10^{-1}

d 6.085×10^2 h 1.25×10^{-3}

3   The distance from the Earth to the Moon is approximately 384 000 000 metres.

Write this number in standard index form.

4   Use your calculator to calculate the following.

a $(3 \times 10^5) \times (3 \times 10^7)$ h $\dfrac{3.9 \times 10^8}{1.3 \times 10^{-5}}$

b $(5 \times 10^5) \times (3.2 \times 10^9)$ i $(2.2 \times 10^2) \div (4.4 \times 10^{11})$

c $(2.4 \times 10^5) \times (3.5 \times 10^7)$ j $1 \div (2.5 \times 10^8)$

d $(4.55 \times 10^5) \times (6.2 \times 10^7)$ k $(5 \times 10^5) + (3 \times 10^6)$

e $(1.5 \times 10^7)^2$ l $(8 \times 10^2) + (8 \times 10^4)$

f $(5 \times 10^{-4})^2$ m $(5.2 \times 10^4) - (5.2 \times 10^3)$

g $\dfrac{8 \times 10^{11}}{4 \times 10^3}$

5   Write the following numbers in order starting with the lowest.

4.8×10^{-6} 4800 4.8×10^4 4.8×10^{-3} 0.00048   480 000

6   A rectangle has length 1.4×10^4 metres and width 2.2×10^3 metres.

Calculate the area and perimeter of the rectangle.

Give your answers in standard index form.

7   The distance to the edge of the observable universe is approximately 4.4×10^{26} metres.

Write this distance in kilometres, giving your answer in standard index form.

8   The speed of light is approximately 3.0×10^8 m/s.

How far will light travel in one week?

Give your answer in standard index form.

! 9   One million in standard index form is 1×10^6.
One billion in standard index form is 1×10^9.
One trillion in standard index form is 1×10^{12}.

a   Write down one billion as an ordinary number.

b   Divide one trillion by one billion. Give your answer in standard index form.

c   Subtract 10 million from 2 billion. Give your answer in standard index form.

10   Anil saves some images onto a memory stick. Each image requires 32 000 bytes of memory. How many images can he save if the memory stick has a memory of 1.36×10^8 bytes?

Give your answer in standard index form.

11   The thickness of a ream of paper (500 sheets) is 4.8 cm.

Work out the thickness of one sheet of paper in millimetres.

Give your answer in standard index form.

2 Assess (k!)

D
C

1   Write in index notation:

     a $10 \times 10 \times 10$ **d** $\frac{1}{10}$

     b $10 \times 10 \times 10 \times 10 \times 10 \times 10 \times 10 \times 10$ **e** $\frac{1}{1000}$

     c 10 **f** 0.0000001

C

2   Simplify the following, leaving your answer in index form.

     a $10^{11} \times 10^6$ **d** $10^{10} \div 10^4$ **g** $(10^3)^7$

     b $10^{14} \times 10^{-5}$ **e** $10^7 \div 10^7$ **h** $(10^4)^5$

     c $\frac{10^8}{10^3}$ **f** $10^3 \div 10^8$ **i** $(10^{-4})^2$

B

3   **a**   The table below shows the areas of some oceans and seas in square kilometres.

Ocean or sea	Area
Arctic Ocean	1.4×10^7 km²
Atlantic Ocean	8.24×10^7 km²
Pacific Ocean	1.65×10^8 km²
Mediterranean Sea	2.50×10^6 km²
Gulf of Mexico	1.54×10^6 km²

     a   Write each area as an ordinary number.

     b   Write a list of the oceans and seas in order of size, starting with the smallest.

4   **a**   Which is bigger: 1.1×10^8 or $99\,999\,999$?

     b   Work out the values of a, b and c in each of the following.

         i $3.5 \times 10^a = 350\,000$

         ii $5.69 \times 10^b = 56.9$

         iii $4.006 \times 10^c = 400\,600$

5   Use your calculator to work out 0.5^6

     Give your answer in standard index form.

6   Each year the UK disposes of approximately 12 billion carrier bags.

     Write this number in standard index form.

7   **a**   Write the following numbers in standard index form.

         i 0.003 **ii** $0.000\,006\,55$ **iii** 0.1

     b   Write as ordinary numbers:

         i 1×10^{-9} **ii** 4.22×10^{-6} **iii** 3.9958×10^{-5}

     c   Work out the value of $(3.52 \times 10^4) \times (2.2 \times 10^{-3})$

     d   Calculate the value of $(3.52 \times 10^4) \div (2.2 \times 10^{-3})$

8   **a**   Work out $(3.28 \times 10^7) \times (9.67 \times 10^{-3})$

b   Work out $(8.39 \times 10^4) \div (1.76 \times 10^{-8})$

c   Work out $(6.25 \times 10^5)^{-3}$

Give your answers in standard index form to three significant figures.

9   A company employs 4.7×10^3 workers. On average, the workers use 2.3×10^2 litres of water per year.

How many litres of water does the company use in a year?

10   In 1900 the world's population was approximately 1.65×10^9.
In 1999 the world's population was approximately 5.98×10^9.

How many times greater was the population in 1999 than the population in 1900?

11   The mass of a hydrogen molecule is 3.3×10^{-24} g.
One litre of hydrogen contains 2.5×10^{22} molecules.

What is the mass of one litre of hydrogen?

12   The Earth is approximately a sphere of radius 6400 km.

a   Use the formula $V = \frac{4}{3}\pi r^3$ to calculate the approximate volume of the Earth.
Give your answer in standard index form in **cubic metres**.

b   The mass of the Earth is 5.97×10^{24} kg.

Use the formula:

$$\text{density} = \frac{\text{mass}}{\text{volume}}$$

to calculate the approximate density of the Earth.

Give your answer in standard index form.

AQA Examination-style questions 🔢

1   **a**   Explain why 36×10^{18} is **not** in standard index form.   *(1 mark)*

   b   The mass of Saturn is 5.7×10^{26} kilograms.
   The mass of Uranus is 8.7×10^{25} kilograms.
   Saturn is heavier than Uranus.

   How many times heavier?
   Give your answer to an appropriate degree of accuracy.   *(3 marks)*

   AQA 2008

Collecting data

Did you know?

The Great Fire of London

Q: How do you know what has happened in the past and what is happening now?

A: This is because someone has recorded it. They have then written about it, talked about it, filmed it or collected data about it.

This has always been the case. You may have heard of the Great Fire of London from 1666. This is probably due to a man called Samuel Pepys. He kept a diary for many decades about life in London. Below is an extract from 2 September 1666.

'So down [I went], with my heart full of trouble, to the Lieutenant of the Tower, who tells me that it begun this morning in the King's baker's house in Pudding Lane, and that it hath burned St. Magnus's Church and most part of Fish Street already. So I [went] down to the waterside, … and there saw a lamentable fire. … Everybody endeavouring to remove their goods, and flinging into the river or bringing them into lighters that layoff; poor people staying in their houses as long as till the very fire touched them, and then running into boats, or clambering from one pair of stairs by the waterside to another. And among other things, the poor pigeons, I perceive, were loth to leave their houses, but hovered about the windows and balconies, till they some of them burned their wings and fell down.'

(The Diary of Samuel Pepys, 1666)

When producing statistics you need to think about how data can be collected, recorded and sorted.

You should already know:

✔ how to calculate with fractions

✔ how to use five bar gates for tallying

✔ how to design and use tally charts and frequency tables for discrete and grouped data.

Key terms

hypothesis	discrete data	survey	controlled experiment	census
raw data	continuous data	open questions	data logging	random sampling
primary data	population	closed questions	data collection sheet	stratified (random)
secondary data	sample	pilot survey	observation sheet	sampling
qualitative data	sample size	observation	two-way table	
quantitative data	questionnaire			

Learn... 3.1 Types of data

The data handling cycle is the framework for work in statistics. It has four stages.

The data-handling cycle

Stage 1
Specify the question
What are you trying to find out?
This leads to your hypothesis.

Stage 2
Collect the data
What data do you need?
How and where will
you collect them?

Stage 3
Process and represent the data
Calculate statistics and use
diagrams to represent data.

Stage 4
Interpret and discuss
What does your data tell you?
Have you answered your
question? Do you have enough
data to answer it? You may
need to pose a new question
and begin the cycle again.

Evaluate

In any statistical project it is usual to go through the data handling cycle at least once.

The first stage is to decide on what you are trying to find out. This leads to the **hypothesis**, a statement that you want to investigate.

The second stage is to think about what data you need and how to collect it.

The third stage is to make calculations and summarise the collected data using tables and diagrams.

The fourth stage involves interpreting the diagrams and calculations you have produced. This should lead to an indication of whether the hypothesis has been supported or not.

After completing the full cycle, it may be necessary to refine the original hypothesis and begin the cycle again.

The way you collect the data, and how you represent them, may depend on the type of data you want.

When data are first collected they are called **raw data**. Raw data are data before they have been sorted.

Data can be **primary data** or **secondary data**.
Primary data are data that are collected to investigate the hypothesis.
Secondary data are data that have already been collected, usually for another purpose.

Data can be **qualitative** or **quantitative**.
Qualitative data are not numerical. These data measure a quality such as taste or colour.
Quantitative data involves numbers of some kind (a quantity).

Quantitative data can be **discrete** or **continuous**.
Discrete data means exact values such as the numbers of people in cars. They are numerical data that can only take certain values.
Continuous data are numerical data that can take any value.
They are always measurements, such as distance or time, which have to be rounded to be recorded.

Link

This chapter will look at stages 1 and 2 of the data handling cycle. Later chapters look at stages 3 and 4.

Example: Draw a table and tick the correct boxes to show whether the following data are qualitative or quantitative, discrete or continuous, and primary or secondary.

   a   Sammi collects information about hair colour from the internet.

   b   Kaye measures the height of 100 people.

   c   Ashad spots the numbers on the sides of trains in the station.

Solution:

Person	Qualitative	Quantitative	Discrete	Continuous	Primary	Secondary
Sammi	✓		✓			✓
Kaye		✓		✓	✓	
Ashad		✓	✓		✓	

AQA *Examiner's tip*

These words and what they mean often appear in questions so you need to learn them!

Data are collected to answer questions.

For example, how many miles can a Formula 1 racing car run on one set of tyres?

You do not need to wait and see, others will have collected data on this.

All the tyres of the same type should run for about the same number of miles.

The **population** of tyres is all the tyres that are the same type.

At some point a **sample** of the tyres will have been tested.

A sample is a small part of the population.

Information about the sample should be true for the population.

The sample size is important.
This is the number of people or items in the sample.

It is important to think carefully about a sample size.

The bigger the sample then the more reliable the information.

So, the more tyres in the sample the more reliable the information.

However, it can be expensive or time consuming to collect data on a very large sample.
You can't sell tyres you have used for testing!

Populations are not always this large. For example, you might be collecting data just in your own school. In this case, the population would be much smaller.

Example: Ellie is investigating the question 'What is the average height of a Year 10 girl?'

    a What is the population for her question?

    b Give an advantage of a large sample.

    c Give a disadvantage of a large sample.

Solution: **a** All Year 10 girls (in the world).

    b The larger the sample the more reliable the results.

    c It would be almost impossible to include all the Year 10 girls across the world. Trying to get too many results will be very time consuming.

Practise...   3.1 Types of data

D C B A A*

1 Copy the table and tick the correct boxes for these data.

 a Nat finds out the cost of a cruise holiday in the newspaper.

 b Prita counts the number of red jelly babies in 100 bags.

 c Niles records the weather at his home every day for one month.

Person	Qualitative	Quantitative	Discrete	Continuous	Primary	Secondary
Nat						
Prita						
Niles						

2 For each of the following say whether the data are discrete or continuous.

 a The number of votes for a party at a general election

 b The number of beans in a tin

 c The weight of recycling each household produces each week

 d How many people watch the ten o'clock news

 e How long it takes to walk to school

 f The number of sheep Farmer Angus has

 g The weights of Farmer Angus' sheep

 h The heights of Year 10 students in your school

 i The number of eggs laid by a hen

3 **a** What is a sample?

 b Why are samples taken rather than looking at the whole population?

 c Give **two** reasons why a sample should not be too large.

4 Write down some:

 a qualitative data about a jumper someone has knitted for you

 b quantitative discrete data about a local football team

 c quantitative continuous data about the launch of a space rocket

 d primary data you might collect about your maths homework

 e secondary data you might collect about keeping a pet rat.

5 Find a copy of a newspaper or use the internet.

Find 10 facts and figures given in the newspaper or on websites.

List the type of data involved each time.

Use as many words which describe data as possible.

Your choices must include data which uses each word below at least once.

Qualitative	Quantitative	Discrete	Continuous	Primary	Secondary

6 A company produces energy-saving light bulbs.

They claim each bulb uses 90% less energy in its lifetime than ordinary bulbs.

Explain how and why sampling will have been used in testing this claim.

Learn... 3.2 Data collection methods

Writing a good questionnaire

One method of obtaining data is to ask people questions using a **questionnaire**.

Surveys often use questionnaires to find out information.

Questions can be **open** or **closed**.

Open questions allow for any response to be made (see Q1 in the example below).

Closed questions control the responses by using options (see Q2 in the example below).

It is important that questionnaires:

✔ are easy to understand

✔ are short and do not ask for irrelevant information

✔ give option boxes where possible

✔ do not have overlap or omissions in them where options boxes are used

✔ are not biased (such as 'Do you agree that ...?')

✔ avoid asking for personal information unless vital to the survey

✔ are tested before being used to show up errors or problems (this is called a **pilot survey**).

Example: A shop manager wants to know the age of his customers.

He considers using one of these questions in a questionnaire.

> **Q1.** How old are you?   Answer_____

or

> **Q2.** Tick the box that contains your age.
>
>   20 or under   20–40   over 50
>   ☐   ☐   ☐

a   What problems might there be with these questions?

b   Write an improved question.

Solution: **a**   Q1 is an open question so all kinds of different responses (answers) could be given.

e.g. $18\frac{1}{2}$, 45, not telling you, over 50.

This would make the data very difficult to organise.

Also people may not wish to give their exact age.

Q2 is a closed question so people are more likely to answer it. However, the option boxes are badly designed.

For example:

The groups overlap. If you are 20 which box do you tick? Some ages are missing. There is no box for people in their 40s.

The groups are quite wide so details are vague about the ages.

b   Tick the box that contains your age.

   ☐   ☐   ☐   ☐

under 20   20–39   40–59   60 or over

> **AQA** **Examiner's tip**
>
> Always make sure you have covered all possible responses.
> Here this is done using an open-ended final group of 60 or over.

Other methods of collecting data

Surveys (and questionnaires) can be carried out in many ways.

Here are the most common.

Each method has advantages and disadvantages.

Method	Description	Advantages	Disadvantages
Face to face interviews/ telephone surveys	This is the most common method of collecting data and involves asking questions of the interviewee.	Can explain more complex questions if necessary.\n\nThe interviewer is likely to be more consistent when they record the responses.\n\nMore likely to get responses than with postal or email surveys.	Takes a lot of time and can be expensive.\n\nThe interviewer may cause bias by influencing answers.\n\nThe interviewee is more likely to lie or to refuse to answer a question.
Postal or email surveys	These surveys involve people being selected and sent a questionnaire.	The interviewee can take their time answering and give more thought to the answer.\n\nInterviewer bias is avoided.\n\nThe cost is usually low.	Low response rates which may cause bias.\n\nCan take a long time.\n\nDifferent people might interpret questions in different ways.
Observation	This means observing the situation directly. For example, counting cars at a motorway junction or observing someone to see what shopping they buy. It can take place over a short or a long period of time.	Usually can be relied upon as those being watched do not know they are being observed and so act naturally.\n\nOften has little cost involved.	The interviewee may react differently because they are being observed.\n\nTakes a lot of time.\n\nOutside influences can affect the observations.\n\nDifferent observers can view the same thing but record it differently.
Controlled experiment	An experiment is more general than you might think and is not just for science. For example, timing cars along a particular piece of road is an experiment. A controlled experiment is where the conditions are stated and are not normally changed.	Results should be reliable.\n\nRepeats of the same experiment are possible if more data is needed.	Getting the right conditions for the experiment may be difficult, costly or time consuming.\n\nThe experiment may need special equipment or expertise.
Data logging	A 'dumb' machine collects data automatically. For example in a shop or car park entrance. The machine could then prevent more cars trying to enter an already full car park.	Once set up, machines can work without needing human resources.\n\nData collection is continued for as long as required.	The machines can break down.\n\nThe data are basic, e.g. you cannot tell if people entering a shop are male, female, adults or children just by using the logged data.

Practise...   3.2 Data collection methods   D C B A A*

D

1   The following questions are taken from different surveys.

Write down one criticism of each question.

Rewrite the question in a more suitable form.

a   How many hours of TV do you watch each week?

Less than one hour ☐   More than one hour ☐

b   What is your favourite football team?

Real Madrid ☐   Luton Town ☐

c   How do you spend your leisure time? (You can only tick one box.)

Doing homework ☐   Playing sport ☐   Reading ☐

Computer games ☐   On the internet ☐   Sleeping ☐

d   You do like football, don't you?

Yes ☐   No ☐

e   How much do you earn each year?

Less than £10 000 ☐   £10 000 to £20 000 ☐   More than £20 000 ☐

f   How often do you go to the cinema?

Rarely ☐   Sometimes ☐   Often ☐

g   Do you or do you not travel by taxi?

Yes ☐   No ☐

h   I hate dogs. What do you think?

So do I ☐   They are OK ☐   Not sure ☐

2   Write down the data collection method in each of these situations.

a   A machine counts entry to a nightclub to prevent it becoming overcrowded.

b   Jez fills in some questions on his PC about his mobile phone contract.

c   Doctor Jekyll records blood pressure rates of people watching horror films.

d   Annie is stopped by a person with a clipboard on the High Street asking about perfume.

e   Iona records where students sit in a classroom the first time they enter it.

C

3   Look again at each of the situations in Question 2.

a   For each, give one advantage of collecting data in the given way.

b   For each, give one disadvantage of collecting data in the given way.

4   Give **two** reasons why a pilot survey might be carried out.

! 5   For each of the following situations write a short questionnaire (of up to four questions) using:

a   closed questions only   **b**   open questions only.

   i   To find out whether an adult is married or not

   ii   To find out the favourite holiday destination of families

   iii   To find out how many hours sleep the average person gets

   iv   To find the most popular pet for pensioners.

c   Explain, for each situation, whether your open or closed questions are better for finding out the desired information.

6 Write a questionnaire which could be used to find out:

a where students have been on holiday in the last two years

b who likes Wayne Rooney

c the cost of newspapers bought by students' families.

Learn...   3.3 Organising data

Sometimes when collecting data you need to design a **data collection sheet** or **observation sheet**.

These can be very simple. The key issue is that any possible item seen can be recorded.

Example: Quinlan is collecting data about the types of vehicles passing his house.

He wants to see if there are differences between weekdays and weekends.

Design an observation sheet that Quinlan could use.

Solution: Here is one possible answer.

	Car	Bus	Lorry	Bicycle	Other
Weekday					
Weekend					

The table in the example above is an example of a **two-way table**.

Two-way tables are used to show more than one aspect of the data at the same time (time of week and type of vehicle).

Two-way tables can show lots of information at once.

Example: Students in a school were asked whether they had school dinners or packed lunches.
Their results are shown in the table.

Write down nine facts that can be obtained from this two-way table.

	Boys	Girls
School dinner	24	16
Packed lunch	12	32

Solution:
1 24 boys have school dinner.

2 16 girls have school dinner.

3 12 boys have packed lunch.

4 32 girls have packed lunch.

5 40 (24 + 16) students have school dinner.

6 44 (12 + 32) students have packed lunch.

7 36 (24 + 12) boys have either school dinner or packed lunch.

8 48 (16 + 32) girls have either school dinner or packed lunch.

9 84 (24 + 16 + 12 + 32) students have either school dinner or packed lunch.

Practise... 3.3 Organising data

D

1 The two-way table shows information about gender and wearing glasses.

	Boys	Girls
Glasses	8	17
No glasses	15	24

Use the table to answer the following questions.

a How many people wear glasses?

b How many girls were in the survey?

c How many boys do not wear glasses?

d What method of data collection could have been used to obtain this data?

2 The table shows the different animals on a farm.

	Sheep	Cattle	Pigs
Male	80		90
Female		70	

The farmer has:
- 130 sheep in total
- 340 male animals
- 600 animals in total.

Copy and complete the table.

3 **a** Design an observation sheet to collect data in each of the following situations.

    i The favourite fruit of Year 10 girls and Year 10 boys

    ii The hair colour of males and females entering a club

    iii How full buses are in the morning and evening rush hours

    b What would be the difficulties in actually recording these data in each of the situations above?

4 Mike thinks the weather is often better in the morning than the afternoon.

Design an observation sheet to collect data to investigate this.

5 The two-way table shows the price of holidays.

Prices per person per week for Costa Packet

	7th April to 5th June	6th June to 21st July	22nd July to 5th Sept
Adult	£124	£168	£215
Child (6–16 years)	£89	£120	£199
Child (0–5 years)	Free	£12	£50

The Brown family consists of two adults and two children aged 3 and 12 years.

They have a maximum of £500 to spend on a one week holiday at Costa Packet.

a On which dates could they go on their holiday?

b Mr Brown says that if they save up another £200 they could have a two week holiday at Costa Packet. Is he correct?

 **Learn... 3.4 Sampling methods**

Whatever method you use to collect data you need to consider sampling.
Sampling is obtaining data from part of a population rather than all of it.

Collecting data from the whole population is called a **census**. This can be very expensive and time consuming.

There are a number of different ways to choose a sample.

Random sampling

For a **random sample** to be taken, every member of the population must have an equal chance of being included in the sample.

The two most common ways of obtaining random samples are:

Use of random numbers

1 Number everyone in the population.

2 Obtain random numbers from a list or a calculator.

3 Match the people on the list with the numbers ignoring repeats.

Use of a 'hat'

1 Write everyone in the population on a separate piece of paper.

2 Put them all in a 'hat'.

3 Draw the required number out without replacing them.

And the winner is...

Example: Paul and Sally decide to have a holiday by choosing two destinations at random from a list of places they like.

Their list is:

Aberdeen	France	London	Tenerife	The list does not need to be in alphabetical order like this, but it can help.
Blackpool	Germany	Majorca	USA	
Chester	Holland	Newquay		
Durham	Italy	Portugal		
Edinburgh	Jamaica	Spain		

a   Describe two different methods of obtaining the random sample.

b   Which is the easier method to use? Explain your answer.

Solution: **a**   Paul and Sally could use random numbers or a 'hat'.

Random numbers: Number the places, e.g.   01 Aberdeen   02 Blackpool   03 Chester   and so on ... up to 17 USA

They need to get 2-digit random numbers as there are more than 10 items.

e.g. 34   76   12   78   32   19   12   55   06

Matching the numbers to the places their random selection would be:

34   Number too large so ignored   19   Number too large so ignored

76   Number too large so ignored   12   Number is a repeat so ignored

12   Choose Majorca   55   Number too large so ignored

78   Number too large so ignored   06   Choose France

32   Number too large so ignored

Paul and Sally would holiday in Majorca and France.

Use of a 'hat': Write each place name on a separate piece of paper.

They could then put the pieces of paper in a container.

Two pieces of paper could then be drawn out.

The names on the two pieces would be the destinations for their holiday.

b   In this case as the 'population' was so small it was far easier to use pieces of paper.

If there are hundreds or thousands of items you should use random numbers.

Stratified random sampling

Most data have some natural groups or strata within them.

For example, people are either men or women.

To get the best possible sample it is often useful to reflect any strata within the population.

So if a college is 70% male students, a sample should be 70% male to be as representative as possible.

For a **stratified random sample**:

- each group is represented by the same proportion as in the population. This allows you to calculate the *number* of items or people
- each item or person should be chosen at random from the group.

Example: Three quarters of a cricket club are non-playing members. The rest are playing members.

How would a stratified random sample of 40 members be chosen?

Solution: The sample of 40 should have the same proportion of playng members and non-playing members as the population.

There should be $\frac{3}{4} \times 40 = 30$ non-playing members in the sample.

There should be $\frac{1}{4} \times 40 = 10$ playing members (or $40 - 30$) in the sample.

The 30 non-playing members should then be chosen at random from all the non-playing members.

The 10 playing members should be chosen at random from all the playing members.

Example: The table shows the number of students in a school by year group.

Year	7	8	9	10	11
Number	200	200	240	220	140

a In a sample of 50 students from this school, stratified by year group, how many of each year group is there?

b Explain how you would then choose the students from Year 7.

Solution: a The proportion in each year group in the sample must be the same as in the population.

Step 1 Add up to find the total $200 + 200 + 240 + 220 + 140 = 1000$

Step 2 Find the fraction of this total for each year e.g. for Year 7 $= \frac{200}{1000}$

Step 3 Calculate the correct fraction of the sample for each year

number in Y7

required sample size

e.g. for Year 7 $\dfrac{200}{1000} \times 50 = 10$

total population

So, 10 Year 7 students would be in the sample.

The table shows the remainder of the calculations.

Year	7	8	9	10	11
Number	200	200	240	220	140
Fraction	$\frac{200}{1000}$	$\frac{200}{1000}$	$\frac{240}{1000}$	$\frac{220}{1000}$	$\frac{140}{1000}$
For a sample size of 50	$\frac{200}{1000} \times 50$ $= 10$ students	$\frac{200}{1000} \times 50$ $= 10$ students	$\frac{240}{1000} \times 50$ $= 12$ students	$\frac{220}{1000} \times 50$ $= 11$ students	$\frac{140}{1000} \times 50$ $= 7$ students

Alternative solution to part a

The population is $200 + 200 + 240 + 220 + 140 = 1000$

The sample required is 50.

The sample fraction is $\frac{50}{1000} = \frac{1}{20}$

This means you just need $\frac{1}{20}$ (5%) of each year total for your sample.

For Year 7,   5% of 200 = 10   For Year 9,   5% of 240 = 12   For Year 11,   5% of 140 = 7

For Year 8,   5% of 200 = 10   For Year 10,   5% of 220 = 11

b   Each set of students should now be chosen by random sampling. For Year 7, number the students from 000 to 199 (or 001 to 200). Calculators can give 3-digit random numbers or tables can be used.

Here is a set of 3-digit random numbers from a random number table.

546   322   108   232   081   002   826   and so on

So far students numbered 108, 081 and 002 are chosen.

The system is continued until 10 students have been chosen.

Remember all numbers too large and any repeats are ignored.

Example: In a company there are:

426 workers on the shop floor

49 supervisors

11 managers.

In a sample of 37 workers, stratified by worker type, how many workers from each type are there?

Solution: Sometimes the numbers for the final sample do not work out nicely.

Step 1   Find the total $426 + 49 + 11 = 486$

Step 2   Fractions of each worker type are:

Shop floor $= \frac{426}{486}$ Supervisors $= \frac{49}{486}$ Managers $= \frac{11}{486}$

Step 3   Calculate the correct fraction of the sample for each group.

Shop floor $= \frac{426}{486} \times 37 = 32.432\ldots$

Supervisors $= \frac{49}{486} \times 37 = 3.730\ldots$

Managers $= \frac{11}{486} \times 37 = 0.837\ldots$

Step 4   Round each value to the nearest integer.

Shop Floor $= 32$ workers in sample

Supervisors $= 4$ workers in sample

Managers $= 1$ worker in sample

Now check the total sample is the correct size. $32 + 4 + 1 = 37$ which is correct.

Sometimes the total sample is one out. In this case it is necessary to round, but not to the nearest integer.

For example a stratified random sample of 116 for a population split as below.

Type	A	B	C	D
Number of type	204	396	185	743

This would give sample sizes before rounding of:

A = 15.48...　　　B = 30.06...　　　C = 14.04...　　　D = 56.40...

Which, after rounding to the nearest integer gives:

A = 15　　　B = 30　　　C = 14　　　D = 56

Now, $15 + 30 + 14 + 56 = 115$ which is 1 short of the desired sample size of 116.

Looking at the decimals, you might round A up to 16.

This is because the decimal for A was closest to being rounded up initially.

The final sample sizes are A = 16, B = 30, C = 14, D = 56

Practise... 3.4 Sampling methods

Where necessary in this exercise use the random number button on your calculator.

AQA **Examiner's tip**

In an exam you would never be asked to produce your own random numbers. You will be given a list.

A

1 Use random numbers to obtain a random sample of five from these towns and cities.

Arbroath, Bangor, Coventry, Derby, Eastbourne, Fishguard, Glasgow, Holyhead, Immingham, Jarrow, Kingston, Liverpool, Manchester, Norwich, Oxford, Plymouth, Queenshead, Reading, Scunthorpe, Tamworth, Uxbridge, Ventnor, Watford, Exeter, Yeovil.

2 The table shows the number of students in each year group of a college.

Year	10	11	12	13
Number	300	300	220	180

a How many would be in each year group of a sample of 50 stratified by year group?

b How would these students then be chosen?

3 The table shows the number of people employed in a department store.

Occupation	Management	Sales	Security	Office
Number	10	130	25	35

a How many would be from each section for a sample of 40 stratified by occupation?

b Explain in detail how the office staff could then be chosen.

4   Tanya wants a panel of 20 people to represent the views of students at her university.

She wants the panel to be stratified by area of study.

The areas students study at her university are given in the table.

Area of study	Social science	The arts	Science	Sport and leisure
Number	3127	1087	2432	976

Calculate the number from each area who should appear on the panel.

5   James is investigating the hypothesis 'children prefer to play on a computer rather than to play in a park'.

He asks a sample of 100 children from his town.

The age distribution of children from his town is given in the table.

Age group	5–7	8–10	11–13	14–16
Number in age group	77	105	133	111

a   Why is it a good idea to stratify the sample?

b   How else could he have stratified the sample apart from by age?

c   How many from each age group should be in James' sample?

6   Look at the Premier League table either online or in a newspaper.

Use random numbers to obtain a random sample of five football teams from the Premier League.

7   The number of students in each year of a large comprehensive school is given below.

Year	7	8	9	10	11
Number in year	244	221	201	195	182

A sample of 100, stratified by year group, is to be taken.

Calculate the number of students from each year to be in the sample.

8   In a car factory, three machines A, B and C produce components.

Machine A produces 40%, Machine B produces 35% and Machine C produces the rest.

2000 components are to be sampled for checking.

Work out the most effective statistical sampling method to do this. Give all necessary details including how the sample should be obtained.

9   A leisure centre is considering changing its opening times.

The centre has 5894 members; 4004 of these are female.

$\frac{3}{4}$ of female members are adults, 70% of male members are adults.

A sample of 200 is to be taken to comment on these plans.

Complete this table for a sample stratified by gender and age.

	Adults	Children
Male		
Female		

3 Assess

D

1 In a survey, 40 adults are asked if they are left handed or right handed.

	Men	Women
Left handed	5	8
Right handed	19	8

Use the table above to answer the following questions.

a How many men are in the survey?

b How many of the 40 adults are right handed?

c What fraction of those asked are right-handed men?

d What percentage of women asked are left handed?

2 Some teachers are asked to choose their favourite snack, out of chocolate and sweets. Some of the results are shown in this table.

	Chocolate	Sweets
Male	24	
Female	16	

a A total of 50 male teachers are asked and 30 teachers choose sweets. Copy and complete the table.

b How many females are asked?

c How many teachers are asked altogether?

d What fraction of the teachers who prefer chocolate are female?
Give your answer in its simplest form.

e How many males who prefer chocolate should be in a sample of 10 stratified by gender and preference?

3 For each of the following say whether the data are discrete or continuous.

a Ages of people

b Goals scored in a school hockey match

c Shoe sizes of people

d The amount of water consumed by a household

e The viewing figures for a TV programme

f The time it takes to get home

g The number of stars that can be seen in the sky

4 **a** Criticise each of the following questionnaire questions.

    i How many hours of television have you watched in the last two months?

    ii Do you or do you not watch news programmes?

b Criticise each of the following questionnaire questions.
Suggest alternatives to find out the required information.

    i What do you think about our new improved fruit juice?

    ii How much do you earn?

    iii Do you or do you not agree with the new bypass?

    iv Would you prefer to sit in a non-smoking area?

    v How often do you have a shower?

C

5   Briefly explain a good method for collecting data in each of these situations.

   a   The average weight of sheep on a farm with 1000 sheep

   b   The favourite building of people in your town

   c   The average amount of time spent on homework each week by students in your school

   d   The average hand span of students in a school

   e   The views of villagers on a new shopping centre

   f   Information on voting intentions at a general election

   g   The number of people entering a shop in the month of December

6   The owners of a small shop claim to have the cheapest prices for fruit and vegetables in a small town.

   Discuss how this could be be tested by explaining how the full data handling cycle could be used in this investigation.

7   Use random numbers to obtain a random sample of three countries from this list.

Angola	Finland	Kazakhstan	Poland	Uganda
Brazil	Greece	Lebanon	Qatar	Venezuela
Colombia	Hungary	Mexico	Russia	Western Samoa
Denmark	Indonesia	New Zealand	Spain	Yemen
Egypt	Jordan	Oman	Tuvalu	Zambia

A

8   The table shows the number of people employed in a factory.

Occupation	Management	Office	Sales	Shop Floor
Number	10	15	30	145

   a   Explain why a random sample of the employees might not be suitable for asking employees about working conditions.

   b   Calculate the number of people from each section for the following sample sizes, stratified by occupation.

   i   For a sample of 20

   ii   For a sample of 35

AQA Examination-style questions 🔑

1   There are 600 members in a sports club.
A stratified sample, by age, is taken.
The table shows the age grouping of the members.
Some information is given in the table.

Age (years)	10–24	25–44	45–60	61+
Number of members	150			120
Number in sample			22	20

Copy and complete the table.

(4 marks)

AQA 2007

4 Percentages

Objectives

Examiners would normally expect students who get these grades to be able to:

D

compare harder percentages, fractions and decimals

work out more difficult percentages of given quantities

increase or decrease by a given percentage

express one quantity as a percentage of another

C

work out a percentage increase or decrease

B

work out compound interest

understand how to use successive percentages

use a multiplier raised to a power to solve problems involving repeated percentage changes

work out reverse percentage problems.

Try this!

Breakfast bits

Do you like bits in your orange juice?

A survey of British people found that:
- 15% won't touch orange juice with bits in it
- 7% won't have it without bits
- 7% won't eat the bits in marmalade or jam.

We have other loves and hates at breakfast time:
- 25% won't eat cereal that has gone soggy
- 14% demand to have matching cutlery
- 7% insist that the crusts are cut off their toast.

Carry out a survey to find out what students in your class love or hate at breakfast time. Surveys nearly always give results in percentages.

Key terms

percentage	credit
Value Added Tax (VAT)	interest
discount	amount
depreciation	compound interest
deposit	rate
balance	principal

You should already know:

✔ place values in decimals

✔ how to put decimals in order of size

✔ how to simplify fractions

✔ how to change fractions to decimals and vice versa.

Learn... 4.1 Percentages, fractions and decimals

1% (1 per cent) means '1 part out of 100' or 'one hundredth'.

It is equivalent to the fraction $\frac{1}{100}$ and the decimal 0.01. In money it is equivalent to '1p in the £1'.

To write other **percentages** as fractions or decimals, divide by 100.

35 hundredths

The figures move 2 places to the right.

For example, $35\% = \frac{35}{100} = \frac{7}{20}$ or $35\% = 35 \div 100 = 0.35$

Use the fraction key on your calculator to simplify fractions, or the equals key.

To write a decimal or fraction as a percentage, multiply by 100 (the inverse operation).

For example, $\frac{3}{5} = \frac{3}{5} \times 100\% = 60\%$ and $0.7 = 0.7 \times 100\% = 70\%$

The figures move 2 places to the left.

On your calculator press

or enter the fraction $\frac{3}{5}$ then press

Example: Which of these is nearest in size to 37%? $\frac{2}{5}$ $\frac{3}{7}$ $\frac{9}{25}$

> **AQA Examiner's tip**
>
> When you are asked to compare fractions and decimals, change them all to percentages.
> Comparing percentages is easier than comparing fractions.

Solution: Write each fraction as a percentage:

$\frac{2}{5} = \frac{2}{5} \times 100\% = 40\%$ This is 3% more than 37%

$\frac{3}{7} = \frac{3}{7} \times 100\% = 42\frac{6}{7}\%$ This is $5\frac{6}{7}\%$ more than 37%

$\frac{9}{25} = \frac{9}{25} \times 100\% = 36\%$ This is 1% less than 37%

So $\frac{9}{25}$ is the nearest in size to 37%

Alternatively

$2 \div 5 \times 100\% = 40\%$

$3 \div 7 \times 100\% = 42.857\%$

$9 \div 25 \times 100\% = 36\%$

There are many ways to find the percentage of a quantity. Here are a few of them.

28% of $£5 = \frac{28}{100} \times 5$ or $£5 \div 100 \times 28$ or $£5 \times 0.28$

28 hundredths of 5   to find 1%   using a decimal multiplier

Try these on your calculator.
The answer is £1.40.

The most efficient way on a calculator is the decimal multiplier.

To find a percentage of a quantity: divide the percentage by 100 to find the decimal multiplier

Then work out: decimal multiplier × quantity

Example: Find: **a** 15% of £48   **b** $42\frac{1}{2}\%$ of £6.35

Solution: **a** Decimal multiplier $= 15 \div 100 = 0.15$

$0.15 \times £48 = £7.20$

Find the multiplier in your head if you can.

b Decimal multiplier $= 42.5 \div 100 = 0.425$

$0.425 \times £6.35 = £2.70$ to nearest penny

You can use one of the other methods to check:

$£48 \div 100 \times 15 = £7.20$

To check this:

$£6.35 \div 100 \times 42.5 = £2.70$

> **AQA Examiner's tip**
>
> Don't forget to add a zero at the end to complete the pence.

Practise... 4.1 Percentages, fractions and decimals k! D C B A A*

D

1 **a** What fraction lies exactly halfway between 49% and 79%?
Give the answer in its simplest form.

b What decimal lies exactly halfway between $12\frac{1}{2}$% and $\frac{7}{8}$?

2 Which of the following fractions is nearest to 50%?

$\frac{4}{10}$ $\frac{9}{20}$ $\frac{14}{30}$ $\frac{19}{40}$

Show how you decided.

3 Tina says that 34% is less than a third. Is she right? Explain your answer.

4 **a** Copy and complete this table of decimal multipliers.

To find	20%	1%	12%	35%	7%	4%	17.5%	2.5%	125%
Multiply by									

b Use the decimal multipliers from your table to work out:

i 20% of 150   **iv** 35% of £500   **vii** 17.5% of £150

ii 1% of 160   **v** 7% of £20   **viii** 2.5% of £32.50

iii 12% of 320   **vi** 4% of £220   **ix** 125% of £25.60

5 Tom has £2200. He gives 15% to his son and 30% to his daughter.
He keeps the rest.
How much money does Tom keep? You **must** show your working.

6 The population of the UK is approximately 61 million.
About 13% of the population is between 5 and 15 years old.
By 2031 the population is expected to rise to 71 million.
About 12.4% of the population is expected to be between 5 and 15 years old.
How many more children between 5 and 15 years old are there expected to be in 2031?

7 The **Value Added Tax (VAT)** on most goods is 17.5% of their value.
Find the VAT on each of these items.

a

£460 + VAT

b

£58.60 + VAT

c

£75.99 + VAT

8 Gas and electricity companies charge 5% VAT on their bills.
Find the VAT on each of these bills.

a

```
Gas you've used (without VAT)
              = £217.62
```

b

```
Cost of electricity used (without VAT)
              = £113.76
```

9   Jan says 'To find 6% of £18 you just need to multiply it by 0.6.'
Is she correct? Explain your answer.

10   Alec saves 30% of his pocket money. He spends £4.20.

    a   What percentage of his pocket money does Alec spend?

    b   What is 1% of Alec's pocket money?

    c   How much does Alec save?

11   Which is greater: $\frac{2}{3}$ of £1 million or 36% of £1.85 million?

12   A town has 24 375 people on the electoral roll.

In an election 29.6% voted Labour, 31.2% voted Conservative and 19.2% voted for other parties.

    a   How many more people voted Conservative than voted Labour?

    b   What fraction of the people did not vote?

13   When selling houses an estate agent charges 3% on the first £50 000 plus 1% of the amount over £50 000.

Find the estate agent's charge on:

    a   a flat sold for £95 000

    b   a house sold for £489 900.

14   In 2009 a TV set cost £650 **excluding** VAT.

During 2009 the rate of VAT was reduced from 17.5% to 15%

How much would be saved on the cost of the TV set **including** VAT after the VAT rate was reduced?

15   The table shows the percentage of UK adults in each group.
A market research firm wants to interview 500 adults.

	Working age	Retirement age
Men	40.0%	8.6%
Women	36.6%	14.8%

How many adults should they interview from each group?
Explain your answer.

16   Amie, Ben, Carrie, Dave, Emma and Fergus share £200 between them.

Amie gets 10% of the £200. Ben gets 20% of what is left.
Carrie gets 30% of what is left after Amie and Ben take their share.
Dave gets 40% of what is left after Amie, Ben and Carrie take their share.
Emma and Fergus share the remainder equally.

Who gets the most?

17   Jamie and Jin carry out a survey of all the students in their year at school.

    a   Jamie says that there are 78 boys.
       Jin says that 48% are girls.
       How many students are in their year at school?

    b   Jamie reports that $\frac{2}{5}$ of these students walk to school.
       Jin says 4% cycle and 18% come by car.
       The rest use public transport.
       How many use public transport?

Learn...   **4.2** **Increasing or decreasing by a percentage**

There are different ways to increase or decrease an amount by a percentage.

For example, to increase £86 by 25% you could do any of the following.

- Find a quarter of £86, then add it on:

$\frac{1}{2}$ of £86 = £43 so $\frac{1}{4}$ of £86 = £21.50
and the increased amount is £86 + £21.50 = £107.50

- Divide £86 by 100, then multiply by 25 to find 25%, then add it on:

25% of £86 = £86 ÷ 100 × 25 = £21.50
so the increased amount is £21.50 + £86 = £107.50

Can you think of
any other ways?

- Use the multiplier 0.25 to find 25%, then add it on:

25% of £86 = 0.25 × £86 = £21.50 then add £86 to give £107.50

The most efficient way on a calculator is to include the original amount in the multiplier.

In this example, the new amount is 125% (100% + 25%) of the original amount.
So the multiplier is 1.25 and the new amount is £86 × 1.25 = £107.50.
This is very quick to do on a calculator. The method is summarised below:

To increase or decrease by a given percentage:
- work out the new quantity as a % of the original quantity
- divide this by 100 to find the multiplier
- multiply the original quantity by the multiplier.

Work out the multiplier
in your head if you can.

Example: Decrease 75 800 by 12.5%

Solution:
New amount	= 100% − 12.5% = 87.5%
Multiplier	= 0.875
Reduced amount	= 75 800 × 0.875 = 66 325

Hint

Find the decimal multiplier by
dividing by 100.

Here it is: 87.5 ÷ 100 = 0.875

You can use a different method
to check.

12.5% of 75 800 = 75 800 ÷ 100 × 12.5
= 9475
Reduced amount = 75 800 − 9475
= 66 325

Use the ANS key on your
calculator to do this.

Practise...   **4.2** **Increase or decrease by a percentage** D C B A A*

D

1   **a**   Copy and complete these tables.

To increase by	20%	40%	8%	3.5%	12.5%	125%
Multiply by						

To decrease by	20%	40%	8%	3.5%	12.5%	1.25%
Multiply by						

b   Use the decimal multipliers from your table to work out:

   i   £15 increased by 20%

   ii   £16 increased by 40%

   iii   £320 increased by 8%

   iv   £250 increased by 3.5%

   v   £92.40 increased by 12.5%

   vi   £9.60 increased by 125%

   vii £15 decreased by 20%

   viii £16 decreased by 40%

   ix   £320 decreased by 8%

   x   £250 decreased by 3.5%

   xi   £92.40 decreased by 12.5%

   xii £9.60 decreased by 1.25%

2   **a**   Increase 120 m by 5%   **e**   Increase 92 000 by 18%

    b   Increase 70 kg by 30%   **f**   Increase 23.6 km by 2.5%

    c   Decrease 8 miles by 15%   **g**   Decrease 782 by 37.5%

    d   Decrease 62.5 litres by 18%   **h**   Decrease £40 by $8\frac{3}{4}$%

    Check your answers using a different method.

3   **a**   The cost of a rail journey is £78.50.
       What is the new price after a 4% increase?
       Check your answer using a different method.

    b   The normal price for a computer game is £12.99.
       The shop reduces this by 20% in a sale.
       What is the sale price?

    c   A ticket for a pop concert on Friday night costs £40.
       It costs $12\frac{1}{2}$% more for a ticket for Saturday night.
       What is the price of a ticket for Saturday night?

4   The table shows the average prices for different types of
houses in a town.
A building society expects these prices to fall by 6%

Find the expected new price for each type of house.
Give your answers to the nearest £100.

Type of house	Average price
Terraced	£97 584
Semi	£149 526
Detached	£226 318

5   Find the total cost of each of these items.

a
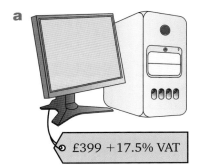
£399 +17.5% VAT

b
```
Cost of electricity
£246.38 + 5% VAT
```

c
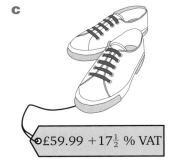
£59.99 +$17\frac{1}{2}$ % VAT

6   The table gives the original prices of some clothes.

The shop reduces these prices by $33\frac{1}{3}$% in a sale.

Find the new prices.

Item	Original price
Trousers	£48.90
Shirt	£29.50
Jumper	£35.95
Gloves	£17.99

7   The cost of going to a theme park was £17.50 last year.
This year it is 2% more.
To find the new price Craig writes down 1.2 × 17.5

    a   What mistake has he made?

    b   Work out the new price correctly.

8   The cost of visiting a zoo is £10 before it puts up its prices by 5%
In the first week after the increase, the zoo gives a **discount** of 5%
Sunita says the cost will be back to the old price.

Do you agree? Explain your answer.

D

C

9 Jasmine says that a 20% increase followed by another 20% increase is the same as a 40% increase.

Is this correct? Explain your answer.

10 A new car costs £12 500. Its value **depreciates** by 8% each year.

How much will it be worth when it is three years old?

> **Bump up your grade**
>
> For Grade C you will need to work the depreciation out for at least two years.

11 The table shows the salaries of some of a company's employees.

The company is discussing these two offers for a pay rise with the employees.

Offer 1
A salary increase of £350 for all employees

Offer 2
A 2% salary increase

Job	Number of employees	Salary (per year)
Clerical assistant	5	£15 400
Factory worker	25	£16 900
Warehouse worker	8	£17 500
Delivery driver	4	£19 750

Compare these offers from the company's and employees' points of view.

12 Abdul wants to buy a TV.

He sees these special offers for the TV in three different shops.

Special Offer
Usual price £589
Sale price 25% off

Buy this TV for
Deposit £100
Balance 12 monthly payments of £35

Spring Sale
Usual price £549
Sale price 20% off

Which offer is cheapest?
Show all working and explain your answer.

Why might Abdul choose one of the other offers?

13 Sally wants to buy a drum kit priced at £495.
She pays a **deposit** of £100.
There are two ways she can pay the rest of the price (the **balance**).

EasyPay Option
7.5% **credit** charge on the balance
6 equal monthly payments

PayLess Option
2.5% **interest** added each month to the amount owed.
Pay £50 per month until the balance is paid off (In the last month Sally will only pay the remaining balance, not a full £50)

Investigate these two options and advise Sally which one is best.

Would your advice be different if EasyPay charged 8% or PayLess charged 1.5% each month?

Learn... 4.3 Successive percentages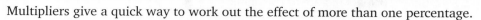

Multipliers give a quick way to work out the effect of more than one percentage.

For example, suppose you know that 56% of the 1250 people at a festival are male.
The multiplier for 56% is 0.56, so the number of males at the festival = 1250 × 0.56 = 700

If you also know that 35% of these males are under 20 years old.   Multiplier for 35%
Then the number of males under 20 years old at the festival = 700 × 0.35 = 245

You can do all of this in one calculation:
The number of males under 20 years old at the festival = 1250 × 0.56 × 0.35 = 245

If you just want to know what percentage of the festival-goers are males under 20 years old, you just need to multiply the multipliers:

0.56 × 0.35 = 0.196 so 19.6% of the festival-goers are males under 20

You can combine percentage changes in the same way.

Example: Jan buys a car for £12 800.
It depreciates by 25% in the first year and 20% in the second year.
What is the car worth after 2 years?

> AQA **Examiner's tip**
> Use multipliers to save time.

Solution: The original price is 100%. In the 1st year the car depreciates by 25%
This means its value goes down from 100% to 75% of the original value.

The multiplier is 0.75

At the end of the 1st year the car is worth £12 800 × 0.75 = £9600

At the start of the 2nd year, the car is worth £9600 ◄— Count this as 100% at start of 2nd year.

After the 2nd year it is worth (100 − 20)% = 80% of this. The multiplier is 0.80

After 2 years the car is worth £9600 × 0.80 = £7680

Check this on your calculator using a single calculation:

Value of car after 2 years = £12 800 × 0.75 × 0.80 = £7680

Multiplier for   Multiplier for
the 1st year   the 2nd year

0.75 × 0.80 = 0.6 = 0.60 so the car is now worth 60% of its original value.
It has lost 40% of its value.

Practise... 4.3 Successive percentages

 1   A school has 1480 students. 20% of the students take part in a survey and 75% of these students say they have the internet at home.

B

  a   How many students took part in the survey?

  b   How many of those surveyed have the internet at home?

  c   Show how you could work out the answer to part **b** in one line using multipliers.

  d   Estimate the number of students at the school who have the internet at home.

B

2   There are 39 375 people at a football match.
24% of these are teenagers. 32% of these teenagers are under 16 years old.
How many teenagers under 16 years old are at the match?

3   Rowan earns £120 for delivering papers.
He spends 5% of his earnings at the newsagents.
45% of his spending at the newsagents is on magazines.
How much does Rowan spend on magazines?

4   A questionnaire about a new speed limit is sent to 2500 villagers.
72% of the questionnaires are returned.
65% of these are in favour of the new speed limit.

How many of the returned questionnaires are in favour of the
new speed limit?

5   A new house costs £95 000 to build and is sold for 26% more.
A year later an estate agent tells the owner that house prices have
gone down by 8%

What is the house worth after the price fall?

6   Val buys shares for £25 000. In the first month they lose 4.8% of their value.
In the following month, the value of the shares increases by 8.5%
What are Val's shares worth at the end of the second month?

7   Ed earns £9.60 per hour. On promotion he gets a pay rise of 5%

After a year in his new job, Ed gets another pay rise of $2\frac{1}{2}$%
How much is Ed paid per hour after the $2\frac{1}{2}$% increase?

8   An insurance company gives Don a 50% discount on the price of his insurance
for making no claims in the past. He gets a further 15% reduction on the
discounted price for agreeing to pay the first £200 on any claim he makes.
What percentage of the original price of the insurance does Don pay?

9   Mr Marks asks, 'Which of these is bigger? An 18% increase followed by a 12%
increase or a 12% increase followed by an 18% increase.'

Katie says that they both give an increase of 30%

a   What mistake has Katie made?

b   What is the correct answer to Mr Marks' question?

10   52% of the adults in a town are women. 86% of these women have jobs and
28% of the women who have jobs work full-time.

What percentage of the town's adult population are:

a   women who work full-time

b   women who work part-time?

11   A bookshop is closing down.
In its final week, books are reduced by
5% more each day as shown in the table.

Monday	Reduced by 10% of usual price
Tuesday	Reduced by 15% of Monday's price
Wednesday	Reduced by 20% of Tuesday's price

a   Pete buys a book at the shop on Wednesday.
What percentage of the usual price does
he pay?

b   Assume the shop continues to reduce prices in this way until it closes on
Saturday. On which day would the price of a book be reduced to less than a
quarter of its usual price?

c   Should Pete have delayed his purchase until Saturday? Explain your answer.

12 A shop buys clothes and usually sells them at a profit of 30%
In a sale it reduces its usual prices by 25%
Does the shop still make a profit? Explain your answer.

 13 A shop usually sells trainers at prices that include a markup of 50% for profit.
The manager decides to have a sale, but still wants to make 20% profit on the
trainers sold.

By what percentage of the usual selling price should the manager reduce the
prices of the trainers for the sale?

 Learn...   4.4 Compound interest

When money is put into a savings account at a bank or building society, interest is paid each year.

Usually the interest is added on to the **amount** already in the account (this is called 'compounding' or paying **compound interest**). For the next year, the amount of money in the account is greater so more interest is paid. This is an example of **repeated proportional change**.

For example, suppose you invest £500 in an account at a fixed **rate** of interest of 4% per year.

At the end of the first year you will have £500 plus 4% of £500, which is 104% of £500.

The multiplier is 1.04

Repeated use of this multiplier gives the amount in the bank after 1 year, 2 years, 3 years and so on. Each calculation can be done in one step on your calculator.

After 1 year, the amount is £500 × 1.04

After 2 years, the amount is £500 × 1.04 × 1.04 = £500 × 1.04²

After 3 years, the amount is £500 × 1.04 × 1.04 × 1.04 = £500 × 1.04³

If left in the account for 10 years, the amount would grow to £500 × 1.04¹⁰

You can work this out on your calculator one year at a time or all at once.

For one year at a time:

Press 5 0 0 = × 1 . 0 4 = then = = = ... (count to 10)   *Using the power key is more efficient.*

To do it all at once, work out 500 × 1.04¹⁰ using the power key.

The amount after 10 years is £740.12, so the interest earned is £740.12 − £500 = £240.12

You can write a formula for the amount in the account after t years.

After every year, the amount in the account is multiplied by 1.04   *An increase of 4% a year for*
So after t years, the amount has been multiplied by 1.04^t *10 years leads to an overall*
increase of about 48%

So the amount in the account after t years is £500 × 1.04^t

Example: A ball falls to the ground from a height of 5 metres.

The height it reaches after each bounce is reduced by 20%

a   Find the height it reaches after the fourth bounce.

b   Show that it will take 8 bounces for the height reached in a bounce to reduce to less than 1 metre.

Solution:

a The multiplier is 0.8 (because the height reached is reduced by 20% to 80% after each bounce).

Height after 4 bounces $= 5 \times 0.8 \times 0.8 \times 0.8 \times 0.8 = 5 \times 0.8^4$
$= 2.048 = 2.05$ metres (to 3 s.f.)

b Height after 7 bounces $= 5 \times 0.8^7$
$= 1.048576$

Height after 8 bounces $= 1.048576 \times 0.8$
$= 0.8388608$

It takes 8 bounces for the height to reduce to less than 1 metre.

> **AQA** *Examiner's tip*
>
> Use this way to save time and avoid errors when you copy amounts.

Practise... 4.4 Compound interest

1 Will invests £600 for two years at 5% compound interest.

Find the amount in the account at the end of two years.

> **Hint**
>
> The initial amount invested is sometimes called the **principal**.

2 £4500 is invested at 4.3% compound interest.

Work out the amount in the account at the end of 3 years.

3 £2000 is invested at 6% compound interest.

    a Write down the amounts in the bank after:

        i 5 years

        ii 10 years.

    b How much interest was earned in:

        i the first 5 years

        ii the second 5 years?

        iii Why is your answer to part **bii** larger than your answer to part **bi**?

4 Work out the compound interest on £16 000 invested for 4 years at $5\frac{3}{4}\%$

5 Carmen puts £2500 into a savings account that pays 4% per year.
She says that after 5 years she will have 20% more.

Is Carmen correct? Explain your answer.

6 Roy wants to invest £5000 for 3 years. He considers two accounts.

Account	Interest Rate	Conditions
Simple Saver	4.6%	Take out the interest at the end of each year
Compound Saver	4.6%	Add the interest to the account each year

Work out how much more interest Roy will get if he uses the Compound Saver.

B

7   Paul buys a car for £7900. Its value depreciates by 20% each year.

   a   Complete a table to show how the car's value reduces over the next 8 years.

   b   Sketch a graph of the car's value against time.

8   A young manager has a job with a starting salary of £19 000 increasing by 2.4% per year.

   a   What is her salary after one year?

   b   What is her salary after five years?

   c   Why may your answer to part **b** not be very realistic?

9   £8000 is invested at $5\frac{1}{2}$% compound interest.

The interest is added at the end of each year. Show that it takes 5 years for the amount to grow to more than £10 000.

10   A report says that the cost of mobile phone calls is falling by 4% per year and that the average cost now is 12 pence per minute. Use this information to estimate how many years it will take for the average cost to fall to:

   a   less than 10 pence per minute

   b   less than 7.5 pence per minute.

11   Rita has earned £16 000 this year. Each year she gets a pay rise of 2%
Rita wonders how long it will be until her yearly earnings reach £20 000.

   a   Show that after 10 pay rises, her yearly earnings will still be less than £20 000.

   b   After how many pay rises will Rita's yearly earnings be more than £20 000?

12   Mike is going to a slimming club. He weighs 112 kg. His target weight is 80 kg.

He aims to lose 3.5% of his body weight per month.
Show that it takes him 10 months to reach his target weight.

⚠ 13   The table gives details of a bank's different savings accounts.
All interest is paid into the accounts after the periods shown.

Account	Interest Rate
Emerald	3.90% paid annually
Ruby	1.94% paid every 6 months
Sapphire	0.96% paid every 3 months

Work out which account gives the best return on £15 000 saved for a year.

⚙ 14   It is thought that about 22 000 polar bears are living in the wild today, but that this number is decreasing by 8% per decade.

Write a short report to illustrate how the number of polar bears is likely to decrease over the next century.

Include calculations and state any assumptions you make.

Investigate the situation for other endangered animals such as the giant panda or black rhino.

❓ 15   £2000 is invested at x% interest over three years.
At the end of three years the money has grown to £2963.
To the nearest percentage find the rate of interest.

Learn... 4.5 Writing one quantity as a percentage of another

To write one quantity as a percentage of another

* Make sure they are in the same units.
* Divide the first quantity by the second. This gives you a decimal (or write the first quantity as a fraction of the second).
* Then multiply by 100. This changes the decimal or fraction to a percentage.

Example: The table shows the marks students get in a test.

a What percentage of students get less than 10?

b Students must get 20 or more to pass.

   i What percentage of students pass?

   ii What percentage of students fail?

Marks	Frequency
0–9	4
10–19	7
20–29	9
30–39	8
40–49	4

Solution: a The total number of students $= 4 + 7 + 9 + 8 + 4 = 32$

4 out of 32 students get less than 10

— divide by the total

Percentage of students who get less than $10 = 4 \div 32 \times 100\% = 12.5\%$

or $\frac{4}{32} \times 100 = 12\frac{1}{2}\%$

b The number of students who get 20 or more $= 9 + 8 + 4 = 21$

   i Percentage of students who pass $= 21 \div 32 \times 100 = 65.6\%$ (to 3 s.f.)

   ii Percentage of students who fail $= 100\% - 65.6\% = 34.4\%$ (to 3 s.f.)

     The number who fail $= 4 + 7 = 11$

     Percentage of students who fail $= \frac{11}{32} \times 100\% = 11 \div 32 = 100\% = 34.4\%$ (to 3 s.f.)

     Note: Check that $65.6\% + 34.4\% = 100\%$

Practise... 4.5 Writing one quantity as a percentage of another

D C B A A*

D

1 Out of 52 people who take a driving test, 34 pass.
What percentage of the people:

   a pass   b fail?

2 It snowed on 4 days in January. What percentage of the days is this?

3 Write:

   a 65p as a percentage of £5.20   e 280 ml as a percentage of 2 litres

   b 5 cm as a percentage of 2 m   f £18 500 as a percentage of £25 000

   c 250 g as a percentage of 5 kg   g 63 000 as a percentage of 4 million

   d 850 mm as a percentage of 5 m   h $1\frac{3}{4}$ of an hour as a percentage of 1 day.

Hint

1 m = 100 cm = 1000 mm   1 kg = 1000 g   1 litre = 1000 ml

4   A school has 6 part-time teachers and 29 full-time teachers.

What percentage of the teachers are:

a   part-time   **b**   full-time?

5   Ten out of fifteen boys and eight out of seventeen girls in a class have school dinners.

What percentage of the whole class:

a   have school dinners   **b**   do not have school dinners?

6   This is Ian's tally chart. It shows the hair colour of the students in his class.

Colour	Tally	Frequency
Blonde	ⅢⅢ ⅢⅢ	10
Brown	ⅢⅢ ⅢⅢ |||	13
Black	ⅢⅢ	5
Red	||	2

a   Ian says 10% of the class has blonde hair. Do you agree? Explain your answer

b   Find the percentage of the class that has each hair colour.

c   Show how you can check your answer to part **b**.

7   The surface area of the Earth is 510 million km². 149 million km² is land and the rest is water.

What percentage of the Earth's surface is water?

8   A class is asked to vote on whether or not they want to take part in a sponsored swim.

The table shows the results.

a   What percentage of boys want to take part?

b   What percentage of girls want to take part?

c   What percentage of the whole class want to take part?

d   What percentage of those who want to take part are boys?

	Vote	
	Yes	No
Girls	8	6
Boys	10	4

9   The table gives the ages of the people who go on an activity holiday.

a   What percentage of the people are under 20 years old?

People who are 40 or over must pay extra insurance.

b   What percentage of people are 40 or over?

Age	Frequency
0–9	1
10–19	36
20–39	33
40–59	8
60 and over	2

⚠ 10   The table gives the number of people who take part in a survey.

a   What percentage of the people are:

  i   men   **ii**   women?

b   What percentage of the people are:

  i   under 20   **ii**   40 or over?

c   What percentage of the men are under 20?

d   What percentage of those under 20 are men?

Age x	Number of	
	Men	Women
$x < 20$	24	25
$20 \leqslant x < 40$	26	28
$40 \leqslant x < 60$	18	23
$60 \leqslant x$	12	14

11 A national travel survey asked people how often they cycled last year.
Use the data to compare the results given in the table below.

Cycled (during last year)	Number of people aged	
	5–15 years	16 years and over
Once or more per week	1371	1576
At least once, but less than once per week	792	2627
Never	883	13 310

Learn... 4.6 Finding a percentage increase or decrease

To write an increase or decrease as a percentage

- Subtract to find the increase or decrease.
- Divide the increase (or decrease) by the **original** amount or write the increase (or decrease) as a fraction of the original amount.
- Then multiply by 100 to change the decimal or fraction to a percentage.

$$\% \text{ increase} = \frac{\text{increase}}{\text{original amount}} \times 100\%$$

Example: A worker takes $1\frac{1}{4}$ minutes to pack a box.
After training he can do it in 50 seconds.
What is the percentage decrease in time?

Solution: Before training, he takes 75 seconds. ⟵ $1\frac{1}{4} \times 60 = 75$

Decrease in time = 75 − 50 = 25 seconds

Percentage decrease = 25 ÷ 75 × 100 = 33.3% (to 1 d.p.)

or $\frac{25}{75} \times 100 = 33\frac{1}{3}\%$

Example: The table shows how the food consumed has changed in the last 4 years.

Food	Average consumption per person per week	
	4 years ago	Now
Meat	462 g	448 g
Fish	98 g	84 g
Vegetables	1.11 kg	1.14 kg
Fruit	1.17 kg	1.28 kg

> **AQA Examiner's tip**
> You must use the **same units**.

Compare these results.

Solution: The amounts are very different, so use percentages to compare the changes.

Work in grams for the meat and fish, and kilograms for the vegetables and fruit.

Meat Decrease = 462 − 448 = 14   **% decrease** = 14 ÷ 462 × 100 = **3.0%**

Fish Decrease = 98 − 84 = 14   **% decrease** = 14 ÷ 98 × 100 = **14.3%**

Veg Increase = 1.14 − 1.11 = 0.03   **% increase** = 0.03 ÷ 1.11 × 100 = **2.7%**

Fruit Increase = 1.28 − 1.17 = 0.11   **% increase** = 0.11 ÷ 1.17 × 100 = **9.4%**

> **AQA Examiner's tip**
> Remember to divide by the **original** amount.

These percentages have been rounded to one decimal place.

The greatest change was in the amount of fish we eat, down by 14.3%
The amount of meat also went down but by a smaller percentage.

The amount of vegetables and fruit both increased. The biggest percentage increase was in fruit.

Practise...

4.6 Finding a percentage increase or decrease

AQA *Examiner's tip*

When answers are not exact, round them to an appropriate degree of accuracy.

1   This year a school has 984 students.
Last year it had 1023 students.

What is the percentage decrease?

> **Bump up your grade**
>
> For Grade C you must be able to write an increase or decrease as a percentage.

2   The price of a bus ride goes up from 95 pence to £1.05

What is the percentage increase?

3   A company employs fewer employees than 10 years ago.
The number of male workers decreased from 8530 to 5380.
The number of female workers decreased from 3150 to 1420.

a   Compare the percentage decrease in the number of male and female workers.

b   Work out the percentage decrease for all workers.

4   The table gives the usual price and sale price of a computer and printer.

Find the percentage reduction in the price of:

	Usual price	Sale price
Computer	£549	£499
Printer	£89	£59

a   the computer

b   the printer

c   the total price.

5   The rent of Sophie's flat has gone up from £120 to £150 per week.
Sophie works out 30 ÷ 150 × 100.
She says the rent has increased by 20%

> **Bump up your grade**
>
> For Grade C you must be able to explain why something is wrong.

a   What mistake has she made?

b   What is the actual percentage increase in Sophie's rent?

6   A sports club advertises on local TV. The table shows the number of male and female members before and after the advert.

	Before	After
Male	46	51
Female	39	44

Compare the percentage increase in the number of male and female members.
Was the advert more successful with males or females?

C

7 A shopkeeper buys a box of 25 pens for £3. He sells them for 20 pence each.

Find the percentage profit.

B

8 It costs a company £0.75 million to make 50 thousand teapots.
The teapots do not sell well. The company sells them off at £12 each.

Find the percentage loss.

! 9 What is the percentage increase if the value of a painting

a increases by a half of its previous value

b doubles

c goes up to five times its previous value?

! 10 Tanya says that you can make more than 100% profit, but you can't make more than 100% loss.

Is she correct? Explain your answer.

11 The table shows differences in what we drink now and 4 years ago.

Drink	Average consumption per person, per week	
	4 years ago	Now
Fruit juice	280 ml	340 ml
Low calorie soft drinks	442 ml	508 ml
Other soft drinks	1.39 litres	1.18 litres
Beverages (e.g. tea, coffee)	5.6 litres	5.6 litres
Alcoholic drinks	763 ml	772 ml

Compare these results.

12 The table gives the UK population in millions.

Year	1971	1976	1981	1986	1991	1996	2001	2006
Population	55.9	56.2	56.4	56.7	57.4	58.2	59.1	60.6

a In which 5 year interval was the percentage increase the greatest?

b Use the data to estimate what the population will be in:

    i 2011   **ii** 2016   **iii** 2021   **iv** 2026   **v** 2031

Explain your method and any assumptions you make.

Learn... 4.7 Reverse percentages

In a reverse percentage problem you start with the final amount and work back to the original amount.

One way of working this out is by using the **unitary** method.
It is based on finding the amount or cost of **one** unit (hence the name 'unitary').
Here the problem is solved by finding **1%**

For example, a digital camera is priced at £156.72 in a sale. It has been reduced by 20%.
What was the original price and how much is saved?

£156.72 is 80% of the original price (since 100% − 20 % = 80%)

So, 1% of the original price = £156.72 ÷ 80 = £1.959

and 100% of the original price = £1.959 × 100 = £195.90

The original price was £195.90.

The saving is £195.90 − £156.72 = £39.18   (or work out 20% of £195.90)

An alternative method is by using the decimal multiplier.

In Learn 4.2 you started with an original amount and found the final amount by multiplying by the multiplier. To reverse the process, you just need to **divide** by the multiplier.

In this case the sale price is 80% of the original price, so the multiplier is 0.80 or 0.8

sale price = 0.8 × original price so original price = sale price ÷ 0.8
                                          = £156.72 ÷ 0.8 = £195.90 (as before)

Both the unitary and multiplier methods also work when amounts have been increased.

Example: A plumber charges £493.50 for a job.

This includes VAT at $17\frac{1}{2}$%

What was the cost before VAT?

> **AQA** *Examiner's tip*
> Look out for examples like this when you need to work out the *original* amount, rather than the final amount.

Solution: The final price = 100% + 17.5% = 117.5% of the original price.

Unitary method

117.5% of the original price	= £493.50
1% of the original price	= £493.50 ÷ 117.5
	= £4.2
100% of the original price	= £4.2 × 100
Original price	= £420

Multiplier method

1.175 × original price	= £493.50
Original price	= £493.50 ÷ 1.175
Original price	= £420

> **AQA** *Examiner's tip*
> Check that your answer is the correct original price. Do this by increasing it by 17.5%. You should get back to £493.50.

As you can see from the example, the multiplier method is a more efficient and powerful method if you can use it successfully.

Practise... 4.7 Reverse percentages   D C B A A*

1   **a**   A film on DVD is reduced by 20% in a sale. The sale price is £7.16.
Work out the price of the DVD before the sale.

b   The cost of insurance for Doug's mountain bike has gone up by 12% this year.
It now costs £39.20.
What was the price of his insurance last year?

c   The population of an island has risen 8% since last year. The population is now 280 000.
What was it a year ago? Give your answer to a suitable degree of accuracy.

B

2　A department store reduces its prices by 30% in a sale.
The reduced price of a television is £875

a　What was the price before the sale?

b　How much do you save by buying the television in the sale?

3　The number of students from a school who are going to university this year has gone up by 16% to 145. How many went to university last year?

4　A website says that the price of a download is 40% less than the price of a CD.
It charges £3.99 to download an album.
How much does the website say you save by downloading the album instead of buying a CD?

5　Sumira wants to buy a computer.

Compsave Price
£564
including 17.5% VAT

PC Perfect Price
£498
+ 17.5% VAT

In which store would Sumira pay less?

6　Kev buys a car for £3150. The price was reduced by 10% for paying by cash.
Kev says 'I saved £315 by paying by cash.'

a　Explain why Kev is wrong.　　**b**　Work out how much Kev saved.

! 7　A finance company charges 12% interest per annum. Helen borrows a sum of money and 2 years later gives the company £6272 to pay off the loan.

How much did Helen borrow?

8　A charity can reclaim VAT on goods and services that it buys.
The table shows the prices it paid for an extension.

Item	Price including VAT (17.5%)
Building materials	£1739.00
Builder's labour charge	£6345.00
Plumber's bill	£493.50
Electrician's bill	£465.30
Decorating bill	£181.89

Work out the total amount of VAT that the charity can reclaim.

9　Stacey has inherited some money. She wants to invest some of it to help with the cost of going to university. She decides to use a building society account that gives a fixed rate of interest of 6% per annum.

How much does Stacey need to put into the account if she wants it to grow to at least £12 000 at the end of four years?

Give your answer to the nearest £.

4 Assess k!

1 The table shows Carl's marks in two tests.

In which test did Carl do better?
You **must** show your working.

Test	Mark
A	52 out of 80
B	60 out of 100

D

2 Meera asks 250 of the 1246 students in her school what they like to read.

a What percentage of the school population does she ask?

b The table gives her results.

	Number of students
Books	102
Comics	146
Magazines	215
Newspapers	76
Other	85

  i Why is the total more than 250?

  ii What percentage of the students in the survey like to read magazines?

  iii What percentage of the students in the survey like to read newspapers?

3 The table shows the results of a cycling test.

	Boys	Girls
Pass	34	28
Fail	16	12

Compare the percentage of boys who passed with the percentage of girls who passed.

4 The price of an mp3 player is reduced from £69.95 to £55.96 in a sale.
Find the percentage reduction.

C

5 William invests £1500 in a building society for 2 years.
The compound interest rate is 4% per annum.

How much will William have in the building society at the end of the second year?

6 Foollah got 32 out of 60 for her first maths paper. She sits the second paper next week. It is out of 75. Foollah needs to get an overall 60% to pass maths.

	Mark	Total
Paper 1	32	60
Paper 2	?	75

B

What mark does she need to get on the second paper?

B

7 In a survey, 300 students are asked how they travel to school. 70% say they use public transport. Of these, 90% say they travel by bus.

How many students in the survey travel by bus to school?

8 A dish contains 5000 bacteria. The number of bacteria increases by 18% per hour.

a How many bacteria will be in the dish after 6 hours?

b Show that it will take over 13 hours for the number of bacteria to exceed fifty thousand.

9 Jensen's speedometer overestimates his speed by 10%

At what speed is he travelling when his speedometer says 66 mph?

10 The price of a games console is £236 after a reduction of 20%

What was the price before the reduction?

11 The table shows Debbie's English and maths test scores in Years 8 and 9.

	Year 8	Year 9
English	53%	75%
Maths	41%	63%

In which subject has Debbie's test score improved the most?

You **must** show your working.

AQA Examination-style questions

1 John has £2000 to invest.
He sees this advert.

SureFire Investments

Don't see your money go up in smoke!

Double your money in 10 years.

The average annual growth of our
investment account is 7.2%

Will John double his money in ten years with SureFire Investments?
You **must** show your working.

(4 marks)

AQA 2006

5 Ratio and proportion

Objectives

Examiners would normally expect students who get these grades to be able to:

D

use ratio notation, including reduction to its simplest form and its links to fraction notation

divide a quantity in a given ratio

solve simple ratio and proportion problems, such as finding the ratio of teachers to students in a school

C

solve more complex ratio and proportion problems such as sharing money in the ratio of people's ages

solve ratio and proportion problems using the unitary method.

Did you know?

A fair world?

Many African countries do not have good healthcare. In Tanzania, the ratio of doctors to people is 0.02 to 1000. This means one doctor for every fifty thousand people!

People in other countries of the world have better access to a doctor. In Cuba, the ratio is 5.9 to 1000.

In the UK, it is 2.2 to 1000. Think about this next time you are in a surgery waiting room.

Key terms

ratio
unitary ratio
proportion
unitary method

You should already know:

✓ how to add, subtract, multiply and divide simple numbers by hand and all numbers with a calculator

✓ how to simplify fractions by hand and by calculator.

Learn... 5.1 Finding and simplifying ratios

Ratios are a good way of comparing quantities such as the number of teachers in a school and the number of students.

The colon symbol is used to express ratio.

In a school with 50 teachers and 800 students, the teacher : student ratio is 50 : 800

You read 50 : 800 as '50 to 800'.

Ratios can be simplified like fractions.

Each number has been divided by 10.

$$\text{Ratio} = 50 : 800 = 5 : 80 = 1 : 16$$

Each number has been divided by 5.

This is just like simplifying fractions $\dfrac{50}{800} = \dfrac{5}{80} = \dfrac{1}{16}$

$\div 10 \quad \div 5$

$\div 10 \quad \div 5$

Ratios with 1 on one side are called **unitary ratios**.

Remember that you can use your calculator to simplify fractions by pressing the ⊜ button.

The simplest form of the ratio 50 : 800 is 1 : 16

This means there is one teacher for every 16 students, and $\frac{1}{16}$ of a teacher for every student.

The **proportion** of teachers in the school is $\frac{1}{17}$ and the proportion of students is $\frac{16}{17}$

Example: A photo is 15 cm high and 25 cm wide.
What is the ratio of height to width in its simplest form?

Solution: The ratio of height to width is 15 cm : 25 cm = 15 : 25 = 3 : 5 (Dividing both numbers by 5).

Why is it important to keep the ratio of height to width the same when changing the size of photos?

Example: The total charge for a meal is £6.16 including 66p service charge.

   a   What is the ratio of the original meal price to the service charge?

   b   The ratio of the original meal price to service charge is the same for all meals. Another meal costs £7.50. What is the service charge on this meal?

Solution: **a**   The original price is £6.16 − £0.66 = £5.50.

So the ratio of original price to service charge is

£5.50 : 66p = 550p : 66p = 550 : 66

> **AQA Examiner's tip**
>
> The units must be the same in a ratio. In this example, £5.50 was changed to pence to match the 66p.

(Both amounts are changed to pence so that the numbers both represent the same thing. You could change them both to pounds but it is probably easier to work with integers rather than decimal numbers.)

      550 : 66 = 50 : 6 (dividing both numbers by 11)

and   50 : 6   = 25 : 3 (dividing both numbers by 2)

The ratio of the original price to the service charge in its simplest form is 25 : 3

So for every 28 pence paid, 25p is for food and 3p is for service.

      $\frac{25}{28}$ is the proportion that is for food

so $\frac{3}{28}$ is the proportion of the total charge that is for service.

b The ratio of the original price to the service charge in its simplest form is 25 : 3

To find the service charge on a meal costing £7.50, multiply both sides of this ratio by 30. 25 : 3 = 750 : 90, so when the meal costs £7.50, the service charge is 0.90 or 90p.

This works well because 750 is a multiple of 25. But what if the meal had cost £7.85 instead of £7.50?

When the numbers are difficult, use a calculator to find a unitary ratio.

This means writing the ratio in the form 1 : n or n : 1.
The working (in pence) is shown below.

original   service
price : charge

This is the 1 : n form of the ratio. →

÷ 25 (25 : 3) ÷ 25
      (1 : 0.12)
× 785 (785 : 94.2) × 785

Starting from the 1 : 0.12 ratio, you can find the service charge on any meal just by multiplying by the price of the meal in pence.

The service charge on the £7.85 meal is 94p to the nearest penny.

AQA *Examiner's tip*

Remember that working with ratios is all about multiplying and dividing not about adding and subtracting.

Practise... 5.1 Finding and simplifying ratios  D C B A A*

D

1 Write each of these ratios as simply as possible.

a	2 : 4	**e**	2 : 12	**i**	36 : 24	**m**	1.5 : 2.5
b	2 : 6	**f**	2 : 14	**j**	25 : 100	**n**	$\frac{2}{3} : \frac{4}{9}$
c	2 : 8	**g**	36 : 12	**k**	25 : 200	**o**	$2\frac{1}{2} : 7\frac{1}{2}$
d	2 : 10	**h**	24 : 18	**l**	0.3 : 0.8	**p**	20% : 80%

2 Simplify each ratio:

a 50 cm : 1 m   **e** 100 m : 2 km

b 20 min : 1 hour   **f** 330 ml : 1½ litres

c 1 kg : 500 g   **g** 750 mm : 5 m

d 2 litres : 250 ml   **h** 1.6 kg : 300 g

Hint

Remember, 1 m = 100 cm = 1000 mm 1 kg = 1000 g and 1 litre = 1000 ml.

3 **a** Write down three different pairs of numbers that are in the ratio 1 : 2

b Explain how you can tell that two numbers are in the ratio 1 : 2

4 The simplest version of all these ratios is 1 : 3. Fill in the gaps.

a 2 : __   **b** 5 : __   **c** __ : 21   **d** __ : 3600   **e** a : __

5 Four of these ratios are the same. Which four?

1 : 2.5 $2\frac{1}{2} : 5\frac{1}{2}$ 3 : 6   0.2 : 0.5   25 : 55   2 : 5   3 : 7.5

D

6 Pippa writes the three pairs of numbers 6 and 9, 9 and 12, and 12 and 15.

She says these pairs of numbers are all in the same ratio.

Is Pippa right? How do you know?

C

7 On a music download site, a song costs 65p and an album costs £6.50.

Find the ratio of the cost of a song to the cost of an album in its simplest form.

8 A book group has men and women in the ratio 2 : 7

a There are 21 women in the group.
How many men are there?

b Two more men join the group.
How many more women are needed to keep
the ratio the same?

! 9 The numbers a and b are in the ratio 2 : 3

a If a is 4, what is b?

b If b is 12, what is a?

c If a is 1, what is b?

d If b is 1, what is a?

e If a and b add up to 10, what are a and b?

! 10 Invent your own question like the one above and find the answers.

11 **a** A photo is 20 cm high and 30 cm wide.
What is the ratio of width to height in its simplest form?

b Another photo measures 25 cm high and 35 cm wide.
Is the ratio of its width to its height the same as the photo in part **a**?

12 A recipe for pastry needs 50 grams of butter and 100 grams of flour.

a What is the ratio of butter to flour? What is the ratio of flour to butter?

b How much butter is needed for 200 grams of flour?

c How much flour is needed for 30 grams of butter?

d What fraction is the butter's weight of the flour's weight?

13 **a** Find, in their simplest forms, the teacher : student ratios for these schools.

School	Number of teachers	Number of students
School 1	75	1500
School 2	15	240
School 3	22	374
School 4	120	1800
School 5	65	1365

b **i** A school with 50 teachers has the same teacher : student ratio as
School 1. How many students does it have?

ii A school with 2000 students has the same teacher : student ratio as
School 1. How many teachers does it have?

c Which school has the smallest number of students for each teacher?

14 Map scales are often expressed in ratio form, such as 1 : 100 000

Look at some maps (perhaps you can use examples from geography).

a How are the scales of the maps shown?
Write down some examples.

b A scale is written as '2 cm to 1 km'.
Write this scale as a unitary ratio.

c The scale 1 : 100 000 can be written as '1 cm to n km'.
Work out the value of n.

d What distance in real life does 3 cm represent on a 1 : 100 000 map?

> **Hint**
> You will need to use these conversions.
> 100 cm = 1 m
> 1000 m = 1 km

15 Copy and complete the table to show the ingredients needed for 18 choux puffs.

	12 choux puffs	18 choux puffs
Flour	3 ounces	
Water	5 ounces	
Eggs	2	

16 Here is a pattern sequence.

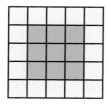
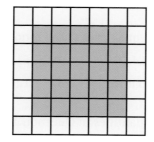

a Does the ratio 'number of green squares : number of yellow squares' increase, decrease or stay the same as the shapes get bigger?
Show how you worked out your answer.

b Draw your own sequence where the ratio of the number of green squares to the number of yellow squares stays the same as the shapes get bigger.

Learn... 5.2 Using ratios to find quantities

You can use ratios to find numbers and quantities.

You can find:
- the number of boys and the number of girls in a school

if you know
- the ratio of boys to girls

and
- the total number of students.

For example, in a school of 1000 students, the ratio of boys to girls is 9 : 11

This means that for every 9 boys there are 11 girls, whatever the size of the school.

9 + 11 = 20, so

9 out of every 20 students are boys so $\frac{9}{20}$ of the students are boys. 11 out of every 20 students are girls so $\frac{11}{20}$ of the students are girls.

The school has 1000 students, so to find the number of boys, work out $\frac{9}{20}$ of 1000.

To find the number of girls work out $\frac{11}{20}$ of 1000.

$\frac{1}{20}$ of 1000 = 1000 ÷ 20 = 50

Number of boys = 50 × 9 = 450

Number of girls = 50 × 11 = 550

> **AQA** *Examiner's tip*
> Check that the number of boys and the number of girls add up to the total number of students in the school.

Example: Jane is 6 years old and Karl is 10 years old.

Their grandmother gives them £24 to share between them in the ratio of their ages.

How much does each child receive?

Solution: The ratio of the Jane's age to Karl's age is $6 : 10 = 3 : 5$

The total number of parts is $3 + 5 = 8$,
so Jane gets $\frac{3}{8}$ of £24 and Karl gets $\frac{5}{8}$ of £24.

$\frac{1}{8}$ of £24 is £24 ÷ 8 = £3

So Jane gets $3 \times \frac{1}{8} = 3 \times £3 = £9$ and
Karl gets $5 \times \frac{1}{8} = 5 \times £3 = £15$

 **AQA**   *Examiner's tip*

Remember to add the numbers in the ratio to find the total number of parts. This is what you have to divide by to find the size of each part.

Practise...   5.2 Using ratios to find quantities   D C B A A*

1   Divide these numbers and quantities in the ratio $1 : 2$

a   150   c   £4.50   e   £1.50

b   300   d   6 litres   f   1.5 litres

2   Divide the numbers and quantities in Question 1 in the following ratios.

a $1 : 4$ b $2 : 3$ c $3 : 7$ d $1 : 2 : 7$

3   In a savings account, the ratio of the amount invested to the interest paid is $50 : 1$

Approximately how much is the interest paid on a savings account that has a total of £10 525 in it?

4   The angles of any pentagon add up to 540 degrees.
The angles of one pentagon are in the ratio $2 : 3 : 4 : 5 : 6$
What is the size of the largest angle?

 5   This table shows the ratio of carbohydrate to fat to protein in some foods.

Food	Carbohydrate : fat : protein
Chicken sandwich	1 : 1 : 1
Grilled salmon	0 : 1 : 1
Yoghurt (whole milk)	1 : 2 : 1
Taco chips	10 : 4 : 1
Bread	7 : 2 : 1
Milk	2 : 3 : 2

a   Work out the amount of fat in 150 g of each of the foods.
Use a calculator for the milk ratios as they do not work out easily.
Round your answers to the nearest 5 grams.

b   Which of these foods would you avoid if you were on a low-fat diet?

c   How many grams of yoghurt would you need to eat to have 100 g of protein?

d   Which of these foods would you avoid if you were on a low-carbohydrate diet?

6 Bronze for coins can be made of copper, tin and zinc in the ratio 95 : 4 : 1

a How much of each metal is needed to make 1 kilogram of bronze?

b How much of each metal is needed to make 10 kilograms of bronze?

c How much of each metal is needed to make half a kilogram of bronze?

d How much zinc would there be in a coin weighing 6 grams?

7 Leena invested £10 000 in a business and Kate invested £3500.

At the end of the year, Leena and Kate share the profits of £70 000 in the ratio of their investments.

How much does each receive?

8 The table shows the number of pupils in five schools together with the ratio of the numbers of boys to the number of girls.

a Which school has the greatest number of boys? Show working to justify your answer.

c Which school has the greatest proportion of boys? Show working to justify your answer.

School	Total number of students	Boy : girl ratio
School A	750	1 : 1
School B	900	4 : 5
School C	1800	4 : 5
School D	1326	6 : 7
School E	1184	301 : 291

Learn... 5.3 The unitary method

You can use the **unitary method** to do all types of percentages as well as ratio and proportion. The method is based on finding the amount or cost of **one** unit (hence the name 'unitary').

So if you know how much 20 litres of petrol cost, you can find the cost of one litre and then the cost of any number of litres.

Example: A teacher pays £27.60 for 6 calculators.

How much does he pay for 15 calculators at the same price each?

Solution: 6 calculators cost £27.60

So 1 calculator costs $\frac{£27.60}{6}$ = £4.60 (divide the cost of 6 calculators by 6)

So 15 calculators cost 15 × £4.60 = £69 (multiply the cost of 1 calculator by 15)

All the calculating can be left to the end if you prefer:

6 calculators cost £27.60

1 calculator costs $\frac{£27.60}{6}$

15 calculators cost 15 × $\frac{£27.60}{6}$ = £69

Dividing by 6 and multiplying by 15 can be done in one step using the

multiplier $\frac{15}{6}$, which is $2\frac{1}{2}$

Cost of 15 calculators = £27.60 × $2\frac{1}{2}$ = £69

This is the same as $\frac{£27.60}{6}$ × 15 or £27.60 × $\frac{15}{6}$

If you feel confident with problems like this, you can do them in one step by combining the multiplication and division, but be careful and check that your answer is sensible.

AQA *Examiner's tip*

Check that your answer is reasonable.

Bump up your grade

To get a Grade C, you should be able to use the unitary method to work out, for example, the cost of 27 items when you know the cost of 5.

Practise... 5.3 The unitary method

C

1 To make chilli con carne for 4 people you need 500 g of beef mince.
How much do you need for 9 people?

2 Katy can type 8 pages in a hour. How long does she take to type 12 pages?

3 Six identical textbooks weigh 1.8 kg. How much do ten of these textbooks weigh?

4 Tom earns £91.80 for 12 hours work. How much does he earn for 15 hours work
at the same rate of pay?

5 A recipe for 12 cheese scones needs 240 g of flour, 60 g of butter and 75 g of cheese.
How much of each ingredient do you need to make 20 cheese scones?

6 Six musicians play a piece of music in 15 minutes. How long do nine musicians
take to play the piece?

7 Abby travelled for three hours on the motorway and covered 190 miles.

   a How far would Abby travel in five hours at the same average speed?

   b How far would she travel in half an hour?

   c How long would it take her to travel 250 miles?

8 Dave drove 246 miles and used 25.4 litres of diesel.

   a How many litres of diesel does Dave need for a 400 mile journey?

   b How far can he go on 10 litres of diesel?

   c What assumption do you have to make to answer these questions?

9 Here are prices for Minty toothpaste.

Size	Amount of toothpaste	Price
Small	50 ml	£0.99
Standard	75 ml	£1.10
Large	100 ml	£1.28

Which size gives the best value for money?
You must show your working.

10 These are prices for different packs of bird seed.

Pack size	Price
5.50 kg	£15.65
12.75 kg	£28.00
25.50 kg	£53.00

   a Find the cost of 1 kg of bird seed for each of the different
pack sizes.

   b Which pack offers best value for money?

   c Find the cost of a 25.50 kg pack if the price per kg was the
same as for the 5.50 kg pack.

   d Give one advantage and one disadvantage of buying a 25.50 kg pack.

11   The weights of objects on other planets are proportional to their weights on Earth. A person weighing 540 newtons on Earth would weigh 90 newtons on the Moon and 1350 newtons on Jupiter.

   a   What would a teenager weighing 360 newtons on Earth weigh on Jupiter?

   b   What would a rock weighing 10 newtons on the Moon weigh on Earth?

   c   What would an astronaut weighing 130 newtons on the Moon weigh on Jupiter?

   d   Express the ratio 'weight of object on Earth : weight of object on Moon : weight of object on Jupiter' in its simplest form.

 12   Sajid worked for 8 hours and was paid £30.

   a   How much will he be paid for working 10 hours at the same rate of pay?

   b   Complete a copy of this table. Plot the values in the table as points on a graph, using the numbers of hours worked as the x-coordinates and the money earned as the corresponding y-coordinates.

Number of hours worked	0	2	4	6	8	10
Money earned (£)						

   c   Explain why the points should lie in a straight line through $(0, 0)$.

   d   Show how to use the graph to find out how much Sajid earns in 5 hours.

? **13**   You may already know about Fibonacci sequences.

   Each term is found by adding together the previous two terms.

   Starting with 1, 1, the series continues:

   1, 1, 2 (as $1 + 1 = 2$)
   1, 1, 2, 3 (as $1 + 2 = 3$)
   1, 1, 2, 3, 5, 8, …

   Carry on the sequence until you have at least 20 terms (you could use a spreadsheet).

   Work out, in the form $1 : n$, the ratio of:

         term 1 to term 2
         term 2 to term 3
         term 3 to term 4
         and so on.

   What do you notice about the ratios as you go through the series?

5

Assess (k!)

D

1   A school has 45 teachers and 810 students.

   a   Express the ratio of teachers to students in its simplest form.

   b   How many teachers would a school of 1200 students need to have the same teacher : student ratio?

2   In a dance class, 30% of the dancers are male.
What is the ratio of male dancers to female dancers?
Give your answer in its simplest form.

3   **a**   Write each of the following ratios in its simplest form.

     i $6 : 8$ **iii** $1000 : 10$ **v** $2\frac{1}{2} : 3\frac{1}{2}$

     ii $27 : 81$ **iv** $\frac{1}{4} : 2$

   b   In a choir there are 12 boys and 18 girls.

     i   Express this as a ratio in its simplest form.

     Two more boys and two more girls join the choir.

     ii   Express the new ratio in its simplest form.

4   To make sugar syrup, 100 grams of sugar is mixed with 250 ml of water.

   a   How many grams of sugar are mixed with 1000 ml (one litre) of water?

   b   How much water is mixed with 150 grams of sugar?

5   Darren gets 16 out of 20 in Test A and 20 out of 25 in Test B.

   a   In which test did he do better?

   b   The next test is marked out of 30. How many marks will Darren need to do as well as he did on Test A?

6   Divide £12 in the ratio:

   a $1 : 2$ **b** $1 : 5$ **c** $1 : 6$

   Explain why part **c** is more difficult than part **a** and part **b**.

7   Jamie is cooking omelettes.

   To make omelettes for 4 people he uses 6 eggs.

   How many eggs does Jamie need to make omelettes for 10 people?

C

8   Gary, Helen and Izzy start a business.
Gary invests £2000, Helen invests £1500 and Izzy invests £2500.
They agree to share any profits in the ratio of their investments.

   In one year the business makes £8100 profit.
How much more does Izzy receive than Helen?

9   The supermarkets 'Lessprice' and 'Lowerpay' both sell packs of pens.

   'Lessprice' sells a pack of 5 pens for £1.25.

   'Lowerpay' sells a pack of 6 pens for £1.44.

   Which supermarket gives the greater value?

10   It takes Kelly 25 seconds to run 200 m. At the same pace, how long will it take her to run:

   a   56 m

   b   128 m?

11   The table shows the approximate population and the number of doctors in some countries of the world.

Country	Population (millions)	Number of doctors
Cuba	10.9	64 300
Israel	5.4	20 600
Italy	57.2	240 000
Nigeria	108.4	30 400
Tanzania	29.7	594
Thailand	58.4	21 600
UK	58.3	128 000
USA	263.6	606 000

   a   In which country is the ratio of doctors : population the greatest?

   b   Work out the number of doctors in each country if the doctors are shared out equally among the total population.

12   A litre of paint covers 15 m² of woodwork.

   a   How much paint is needed for 50 m² of woodwork?

   b   You can buy the paint in the different sizes of tin shown.

      What would you buy to paint 50 m² of woodwork? Explain your answer.

AQA Examination-style questions

1   Year 10 and Year 11 students are in an assembly.
Here are some facts about the students in the assembly.

Year	boys : girls	Student data
10	4 : 5	84 boys
11	2 : 3	150 students

Work out the total number of girls in the assembly.
You **must** show your working.

(5 marks)

AQA 2008

6 Statistical measures

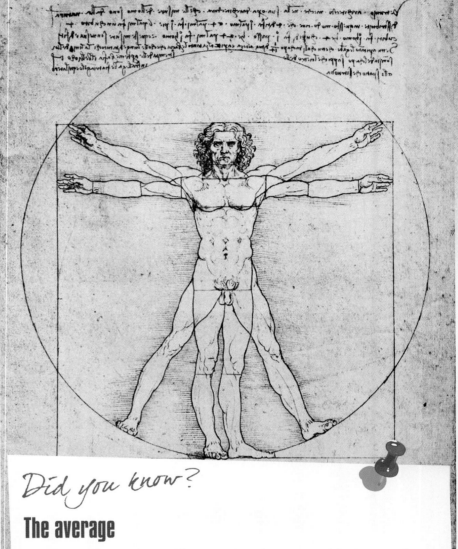

Did you know?

The average

The **Vitruvian Man** is a famous drawing created by Leonardo da Vinci around 1487. His notes offer an explanation of the average man including:

- the length of a man's outspread arms is equal to his height
- the length of a man's foot is one-sixth of a man's height
- the length of the hand is one-tenth of a man's height
- the distance from the elbow to the armpit is one-eighth of a man's height.

Do these measurement still apply to the average man? What about the average woman?

Learn... 6.1 Frequency distributions

A **frequency distribution** shows how often individual values occur (the frequency).

The information is usually shown in a frequency table.

A frequency table shows the values and their frequency.

Frequency distributions are usually used with **discrete data**.

Discrete data are data that can only take individual values.

For example, the number of cars is discrete data. You cannot have 2.3 cars!

This frequency table shows the number of pets for students at a school.

There are 5 students with no pets, 11 students with 1 pet, 8 students with 2 pets, … and so on.

You can use the frequency table to calculate measures of average and measures of spread.

Number of pets (x)	Frequency (f)
0	5
1	11
2	8
3	5
4	2

Example: For the frequency distribution above, find:

a the average

b the spread.

Solution: **a**   The most common measures of **average** are the **mean**, the **mode** and the **median**.

Mean

The mean is the total of all the values divided by the number of values.

The mean here is the total number of pets divided by the total number of students.

$$\text{Mean} = \frac{\text{the total of (frequencies} \times \text{values)}}{\text{the total of frequencies}} = \frac{\Sigma fx}{\Sigma f}$$

where Σ means 'the sum of'.

Number of pets (x)	Frequency (f)	Frequency $\times$ number of pets (fx)
0	5	$0 \times 5 = 0$
1	11	$1 \times 11 = 11$
2	8	$2 \times 8 = 16$
3	5	$3 \times 5 = 15$
4	2	$4 \times 2 = 8$
	$\Sigma f = 5 + 11 + 8 + 5 + 2 = 31$	$\Sigma fx = 0 + 11 + 16 + 15 + 8 = 50$

$$\text{Mean} = \frac{\text{the total of (frequencies} \times \text{values)}}{\text{the total of frequencies}} = \frac{\Sigma fx}{\Sigma f}$$

$$= \frac{50}{31}$$

$$= 1.6129...$$

Mean = 1.6 (to 1 d.p.)

The mean is a useful measure of average.

Mode

The mode is the value which has the highest frequency next to it (i.e. the value occurring most often).

'Number of pets is 1' has a frequency of 11 and all the other frequencies are less than this.

Mode = 1

The mode is the number that occurs most frequently.

Median

The median is the middle value when the data are listed in order.

It does not matter whether you go from smallest to highest or the other way round.

The data have 31 values so the median is the $\left(\frac{31+1}{2}\right)$th value = 16th value. The data are already ordered in the table.

The first 5 values are 0, the next 11 are 1, so the 16th value is 1.

0 0 0 0 0 1 1 1 1 1 1 1 1 1 1 1 2 2 2 ...

16th

Median = 1

The median is the middle value when the data are arranged in order.

The median can also be found using the 'running totals' of the frequencies as follows:

Number of pets (x)	Frequency (f)	Running total
0	5	5
1	11	5 + 11 = 16
2	8	5 + 11 + 8 = 24
3	5	5 + 11 + 8 + 5 = 29
4	2	5 + 11 + 8 + 5 + 2 = 31

The 16th value will lie in this interval so the median is 1.

b The most common measures of spread are the **range** and the **inter-quartile range**.

Range

The range is the difference between the highest value and the lowest value.

The range = 4 − 0 = 4

Range = 4

The range is a measure of how spread out the data are.

Inter-quartile range

The inter-quartile range is the difference between the upper quartile and the lower quartile.

The lower quartile is the $\frac{1}{4}(n+1)$th value = $\frac{1}{4}(31+1)$th value = 8th value

The upper quartile is the $\frac{3}{4}(n+1)$th value = $\frac{3}{4}(31+1)$th value = 24th value

The quartiles can be found using the same method as the median or else you can consider the 'running totals' of the frequencies as follows.

Number of pets (x)	Frequency (f)	Running total
0	5	5
1	11	5 + 11 = 16
2	8	5 + 11 + 8 = 24
3	5	5 + 11 + 8 + 5 = 29
4	2	5 + 11 + 8 + 5 + 2 = 31

The 8th value will lie in this interval so the lower quartile is 1.

The 24th value will lie in this interval so the upper quartile is 2.

The inter-quartile range = 2 − 1 = 1

Inter-quartile range = 1

The inter-quartile range is a measure of how spread out the data are.

It focuses on the middle 50 per cent of the distribution and is not affected by extreme values.

Practise...   6.1 **Frequency distributions**   D C B A A*

1   Andy keeps a record of the points he scores in tennis games.

The table shows results for 40 games he lost.

For the data work out:

a   the mean

b   the median

c   the mode

d   the range

e   the inter-quartile range

f   the percentage of these games where Andy scored 30 points.

Score	Frequency
0	5
15	11
30	20
40	7

D

2   The frequency table shows the speed limit of all the roads in one county.

Speed limit (miles per hour)	20	30	40	50	60	70
Number of roads	8	88	52	23	150	3

a   Work out the number of roads that have a speed limit of:

    i   less than 40 miles per hour

    ii   50 miles per hour or more.

b   What percentage of roads have a speed limit of 30 miles per hour? Give your answer to two decimal places.

c   What is the range of speed limits in this county?

d   Work out the modal speed limit for the county.

e   Work out the median speed limit for the county.

f   Work out the mean speed limit for the county.

D

3 Ryan throws a dice 100 times. The frequency distribution shows his scores.

Score (x)	Frequency (f)	Frequency × score (fx)
1	18	
2	19	
3	16	
4	12	
5	15	
6	20	

AQA *Examiner's tip*

It is possible to use a calculator's statistical functions to find the mean of a frequency distribution. Make sure you are in statistical mode and enter the data value or midpoint followed by its frequency each time.

a Find his mean score.

b Find the median score.

c What is the modal score?

d Work out the range of scores from these 100 dice throws.

e What fraction of the throws resulted in a 6?
Give your answer in its simplest form.

D
C

4 The number of people in a sample of 100 cars is given in the frequency table.

Number of people	1	2	3	4
Frequency	60	32	6	2

a Write down the median number of people in these cars.

b Work out the mean number of people in these cars.

c Which is more useful in predicting the number of people in the next car to come along? Explain your answer.

C

5 This table shows some information about storms in the USA.

	Number of rumbles of thunder
Mean	3.1
Median	3
Mode	3

This table shows the number of rumbles of thunder heard in one minute time periods during a storm in another country.

Number of rumbles of thunder	Frequency
0	11
1	24
2	14
3	3
4	1
5	2

Hint

To compare this storm with storms in the USA you need to work out the average values even though you have **not** been asked to.

Compare this storm with storms in the USA.

⚠ 6 Fill in the frequencies so that the median is 10 and the mode is 9.

x	f
8	
9	
10	
11	

7   Make four copies of this frequency table.

x	f
2	35
4	65
6	19
8	

a   Complete one copy of the table so that x has a mode of 8.

b   Complete one copy of the table so that x has a median of 4.

c   Complete one copy of the table so that x has a median of 5.

d   Complete one copy of the table so that x has a mean of 4.

8   The table shows the number of bedrooms in a sample of houses from a town centre and a village.

Number of bedrooms	Number of houses	
	Town centre	Village
1	8	3
2	19	9
3	6	10
4	2	8
5	0	5

Hint

As well as using measures of average and spread to make the comparisons you could also use percentages.

Compare the number of bedrooms in these two samples.

9   Two bus companies, Super Express and Big Bus, run a bus service between the same two towns along the same route.

George is investigating the punctuality of their buses on this route.

He records the number of minutes late, rounded to the nearest 5 minutes, for a sample of buses from each company over a one week period.

Here are the data George collects for Big Bus.

0	0	0	0	0	0	0	0	0	0
0	0	0	0	0	0	0	0	5	5
5	5	5	5	5	5	10	10	10	10
10	10	10	10	10	10	15	15	25	35

Here are the data George collects for Super Express.

Hint

When you are given sets of raw data, the first thing to do is sort out the data in a frequency table.

0	0	0	0	5	5	5	5	5	5
5	5	5	5	5	5	5	5	5	5
5	5	5	10	10	15	15	15	15	15
15	15	15	15	15	15	15	15	15	15
15	20	20	20	20	20	20	20	20	20

a   Compare the punctuality of the two bus companies.

b   Give a reason why you might be more interested in the range of the times late rather than the average.

10   A frequency distribution for the variable x has the following features.

The total number of pieces of data is 100.

The mean of x is 5.   The mode of x is 3.

The median of x is 4.   The range of x is 7.

Copy and complete this frequency table to show a possible distribution for x.

x	0	1	2	3	4	5	6	7	8	9	10
Frequency											

Learn... 6.2 Grouped frequency distributions

A grouped frequency distribution shows how often **grouped data** values occur (the frequency).

The information is usually shown in a grouped frequency table.

A grouped frequency table shows the values and their frequency.

Grouped frequency distributions are usually used with **continuous data**.

Continuous data are data which can take any numerical value. Length and weight are common examples of continuous data.

Discrete data can only take individual values. Shoe sizes are an example.

You can use the grouped frequency table to calculate measures of average and measures of spread as before.

Mean

The mean is the total of all the values divided by the number of values.

$$\text{Mean} = \frac{\text{the total of (frequencies} \times \text{values)}}{\text{the total of frequencies}} = \frac{\Sigma fx}{\Sigma f}$$

where Σ means 'the sum of'.

As the data are grouped, you will need to use the midpoint of each group to represent the value.

Discrete data	Continuous data
To find the midpoint, add together the largest and smallest values of each group and divide the answer by two.	To find the midpoint, add together the smallest possible value (lower bound) and the largest possible value (upper bound) for each group and divide the answer by two.

Bump up your grade

You need to be able to find the mean of grouped data to get a Grade C.

Mode

The mode is the value which has the highest frequency next to it (i.e. the value occurring most often).

For grouped data it is more usual to find the **modal class**.

The modal class is the class with the highest frequency.

Median

The median is the middle value when the data are listed in order.

For grouped data it is more usual to find the group containing the median.

Graphical work (see Chapter 7) is often used to estimate the median.

Range

The range is the highest value take away lowest value.

For grouped data it is not always possible to identify the highest value and the lowest value. However, it can be estimated as: Highest value in highest group − lowest value in lowest group.

Inter-quartile range

The inter-quartile range is the difference between the upper quartile and the lower quartile.

For grouped data it is not always possible to identify the upper quartile and the lower quartile.

Graphical work (see Chapter 7) is often used to estimate the quartiles.

Example: The table shows the time taken for students to solve a simple puzzle.

$20 \leqslant t < 30$ covers all the values between 20 and 30 seconds. The 20 is included in the group whereas the 30 will be in the $30 \leqslant t < 40$ group. The range of values is called a **class interval**.

Time, t (seconds)	Frequency
$10 \leqslant t < 20$	30
$20 \leqslant t < 30$	35
$30 \leqslant t < 40$	20
$40 \leqslant t < 50$	10
$50 \leqslant t < 60$	5

Use the information in the grouped frequency table to:

a write down the modal class

b work out the class which contains the median

c calculate an estimate of the mean time taken to solve the puzzle.

Solution:

a The modal class is the class with the highest frequency.

This is the class $20 \leqslant t < 30$ (as there are 35 students in this class).

b The median is the middle value when the data are listed in order.

In this case the middle value is the 50th value.

> **Hint**
>
> The median is the $\frac{1}{2}(n + 1)$th value but for large numbers the $\frac{n}{2}$th value is sufficient.

The median can also be found using the 'running totals' of the frequencies as follows.

The 50th value will lie in this interval so the median lies in ⟶ the $20 \leqslant t < 30$ class.

Time, t (seconds)	Frequency	Running total
$10 \leqslant t < 20$	30	30
$20 \leqslant t < 30$	35	$30 + 35 = 65$
$30 \leqslant t < 40$	20	$30 + 35 + 20 = 85$
$40 \leqslant t < 50$	10	$30 + 35 + 20 + 10 = 95$
$50 \leqslant t < 60$	5	$30 + 35 + 20 + 10 + 5 = 100$

The $20 \leqslant t < 30$ class contains the median.

c The mean is the total of all the values divided by the number of values.

$$\text{Mean} = \frac{\text{the total of (frequencies} \times \text{values)}}{\text{the total of frequencies}} = \frac{\Sigma ft}{\Sigma f}$$

where Σ means 'the sum of'.

As the data are grouped, you will need to use the midpoint of each group.

An additional column should be added to the table for the midpoints.

Time, t (seconds)	Frequency (f)	Midpoint (t)	Frequency × midpoint (ft)
$10 \leqslant t < 20$	30	15	$30 \times 15 = 450$
$20 \leqslant t < 30$	35	25	$35 \times 25 = 875$
$30 \leqslant t < 40$	20	35	$20 \times 35 = 700$
$40 \leqslant t < 50$	10	45	$10 \times 45 = 450$
$50 \leqslant t < 60$	5	55	$5 \times 55 = 275$
	$\Sigma f = 100$		$\Sigma ft = 2750$

$$\text{Mean} = \frac{\text{the total of (frequencies} \times \text{values)}}{\text{the total of frequencies}} = \frac{\Sigma ft}{\Sigma f}$$

$$= \frac{2750}{100}$$

$$= 27.5$$

Mean = 27.5 seconds

> **Hint**
>
> Remember that this is only an estimate of the mean as we do not know how the numbers are distributed within each group. Using the midpoint gives an approximation only.

> **AQA** *Examiner's tip*
>
> Remember to check that the answer you have obtained is sensible for the data. Your answer must lie within the range of the data. If it doesn't you have made a mistake.

Practise... 6.2 Grouped frequency distributions  D C B A A*

D
C

1   The table shows the time in minutes people have to wait to be served in a shop.

Time, t (minutes)	Frequency
$0 \leqslant t < 2$	8
$2 \leqslant t < 4$	14
$4 \leqslant t < 6$	6
$6 \leqslant t < 8$	4
$8 \leqslant t < 10$	2

a   Write down the modal class.

b   In which group does the median lie?

c   What percentage of people waited more than 8 minutes?
Give your answer to one significant figure.

d   Calculate an estimate of the mean waiting time.
Explain why your answer is an estimate.

e   Estimate the range of the waiting times.

2   In a survey of reading ability the scores obtained are given in this table.

Reading score	0–4	5–9	10–14	15–19	20–24	25–29	30–34
Frequency	15	60	125	260	250	200	90

What is the modal class?

Calculate an estimate of the mean reading score.

3   The table shows the weights of 10 letters.

Weight, x (grams)	$0 \leqslant x < 20$	$20 \leqslant x < 40$	$40 \leqslant x < 60$	$60 \leqslant x < 80$	$80 \leqslant x < 100$
Number of letters	2	3	2	2	1

Calculate an estimate of the mean weight of a letter.

4   The table shows the weekly wages of 40 staff in a small company.

Wage, x (£)	$50 \leqslant x < 100$	$100 \leqslant x < 150$	$150 \leqslant x < 200$	$200 \leqslant x < 250$	$250 \leqslant x < 300$	$300 \leqslant x < 350$
Frequency	5	13	11	9	0	2

a   Work out:

   i   the modal class

   ii   the class that contains the median

   iii   an estimate of the mean.

b   Which average should you use to compare the wages with another company?
Give a reason for your answer.

5   A company produces three million packets of crisps each day.
It states on each packet that the bag contains 25 g of crisps.
To test this, the crisps in a sample of 1000 bags are weighed.
The results are shown in the table.

Weight, w (grams)	Frequency
$23.5 \leqslant w < 24.5$	20
$24.5 \leqslant w < 25.5$	733
$25.5 \leqslant w < 26.5$	194
$26.5 \leqslant w < 27.5$	53

a   Calculate an estimate of the mean weight of a packet of crisps.

b   Is the company justified in stating that each bag contains 25 g of crisps?

c   What percentage of the packets of crisps weigh under 25 g?

d   Estimate the number of bags weighing under 25 g produced in one day.

6   Two machines are each designed to produce paper 0.3 mm thick.

The frequency distributions below show the actual output of a sample for each machine.

Thickness, t (mm)	Machine A Frequency	Machine B Frequency
$0.27 \leqslant t < 0.28$	2	1
$0.28 \leqslant t < 0.29$	7	50
$0.29 \leqslant t < 0.30$	32	42
$0.30 \leqslant t < 0.31$	50	5
$0.31 \leqslant t < 0.32$	9	2

Compare the outputs of the two machines using suitable calculations.

Which machine is producing paper closer to the required thickness?

7   The table shows the weights of 100 vehicles travelling on a road.

Weight, w (tonnes)	$0 < w \leqslant 1$	$1 < w \leqslant 2$	$2 < w \leqslant 3$	$3 < w \leqslant 4$
Frequency	53	43	5	2

a   Estimate the mean weight of the vehicles on the road.

b   Explain why this value is only an estimate.

8   The frequency distribution shows the time taken by 50 people to complete a supermarket shop.

Time, t (minutes)	$10 \leqslant t < 20$	$20 \leqslant t < 30$	$30 \leqslant t < 40$	$40 \leqslant t < 50$	50 or over
Frequency	13	21	8	5	3

a   Find the class which contains the median length of time.

b   Explain why the class '50 or over' makes estimating the mean time difficult.

c   Yasmin uses the table to estimate the mean length of time spent shopping for these 50 people.

She correctly obtains a value of 28.4

Work out the class interval she used instead of '50 or over'.

6 ## Assess

D

1 a Look at this set of numbers.

1   2   2   3   4   4   4   4   7   7   33   37

How can you tell at a glance that the mean is **larger** than the median?

b Look at this set of numbers.

1   1   22   23   25   25   25   25   26   26   26   27

How can you tell at a glance that the mean is **smaller** than the median?

c Remember that an average is one number that best represents a set of numbers.

In parts **a** and **b** which average, the mean or the median, best represents each set of values?

Give a reason for your answer.

2 a Write down two **different** sets of five numbers that have the **same** median and mode but a different mean.
Work out the range of each set.

b Write down two **different** sets of five numbers that have the same mean and range but a different mode.
Work out the median of each set.

3 Leela rolls an eight-sided dice 200 times. Her results are shown in the frequency table below.

Score	1	2	3	4	5	6	7	8
Frequency	19	26	24	25	22	33	21	30

a Write down the modal score.

b Work out the median score.

c Work out the mean score.

4 Debbie asks students in her class how many brothers and sisters they have.
She puts the information in a table.

```
              5 │  1
              4 │  0   2
Number        3 │  2   2   1
  of          2 │  4   3   3   1
brothers      1 │  5   3   2   1   1
              0 │  5   6   2   1       1
                └──────────────────────────
                   0   1   2   3   4   5
                      Number of sisters
```

a How many people have no sisters?

b How many people have only one brother?

c How many people have equal numbers of brothers and sisters?

d How many people did Debbie survey altogether?

e What is the modal number of sisters?

f Calculate the mean number of brothers.

5 The table shows the number of letters in the words in a game of scrabble.

Numbers of letters in word	2	3	4	5	6	7
Frequency	10	7	5	5	6	3

a Write down the modal word length.

b Work out the median word length.

c Work out the mean word length.

d Which of these averages do you feel is most suitable for these data? Give a reason for your answer.

6 In a survey of the number of people in a household the following information was collected from 50 houses.

Number of people in a household	Number of households
1	9
2	19
3	9
4	8
5	4
6	1
Total	50

a Find the mean, median, mode and range of household sizes.

b Which average is the best one to use to represent the data? Give a reason for your answer.

7 This table gives the number of years' service by 50 teachers at the Clare School.

Number of years' service	0–4	5–9	10–14	15–19	20–24	25–29
Number of teachers	11	15	4	10	6	4

a Find the modal class.

b Calculate an estimate of the mean.

8 The heights achieved by Year 10 high jumpers in a trial are summarised below.

Height, h (cm)	$149.5 \leq h < 154.5$	$154.5 \leq h < 159.5$	$159.5 \leq h < 164.5$	$164.5 \leq h < 169.5$
Frequency	4	21	18	7

a Estimate the mean height jumped in the trial.

b John says 'I won the trial, I jumped 1 metre 65 cm.' Comment on John's statement.

9 The length of 40 political speeches in 1900 in the House of Commons is given in the table.

Length, x (minutes)	$10 \leq x < 20$	$20 \leq x < 30$	$30 \leq x < 40$	$40 \leq x < 50$	$50 \leq x < 60$
Frequency	7	8	16	6	2

By 2008 the mean speech length was 45% longer than 1900.

Estimate the mean speech length for 2008.

C

10 In a science lesson 30 runner bean plants were measured.
Here are the results correct to the nearest centimetre.

6.2	5.4	8.9	12.1	6.5	9.3	7.2	12.7	10.2	5.4
7.7	9.5	11.1	8.6	7.0	13.5	12.7	5.6	15.4	12.3
13.4	9.5	6.7	8.6	9.1	11.5	14.2	13.5	8.8	9.7

The teacher suggested that the data were put into groups.

Length, l (cm)	Tally	Frequency
$5 \leqslant l < 7$		
$7 \leqslant l < 9$		
$9 \leqslant l < 11$		
$11 \leqslant l < 13$		
$13 \leqslant l < 15$		
$15 \leqslant l < 17$		

a Copy and complete the table.

b Use the information to work out an estimate of the mean height of the plants.

c Calculate the mean from the original data.

d Why is your answer to part **b** only an estimate of the mean?

11 The table shows the distribution of three sets of discrete data A, B and C.

	Frequency		
x	Set A	Set B	Set C
0	1	1	0
1	1	2	6
2	7	7	2
3	6	6	10
4	0	4	1

Find as many reasons as you can for each data set to be the odd one out.

Show working to justify your answers.

12 Tom and Sara are investigating this hypothesis about the game of Snakes and Ladders.

'The higher the average score you get when you roll the dice the more likely you are to win.'

Here are the scores in games that Tom won.

Score	Frequency	
	Tom	Sara
1	9	7
2	11	10
3	8	12
4	8	11
5	14	8
6	12	9

Hint

Snakes and Ladders

In a game of Snakes and Ladders you can land on squares that tell you to miss a turn or throw again.

So Tom and Sara are likely to throw the dice a different number of times in a game.

Here are the scores in games that Sara won.

Score	Frequency	
	Tom	Sara
1	11	8
2	8	7
3	11	9
4	7	10
5	8	8
6	10	14

Investigate the hypothesis.

Hint

When you are given sets of data and are asked to investigate a hypothesis you often have to **compare** measures of average and spread.

In this question there are four different sets of data. You have to work out the average score for all four sets. Then you have to decide what comparisons to make that will help you decide whether the hypothesis is true or not.

In this question, you don't have to work out the range. Why?

The best average to use is the mean. Why?

Don't forget to write a conclusion.

AQA Examination-style questions

1   The table shows the heights of 30 students in a class.

Height, h (cm)	Number of students
$140 < h \leqslant 144$	4
$144 < h \leqslant 148$	5
$148 < h \leqslant 152$	8
$152 < h \leqslant 156$	7
$156 < h \leqslant 160$	5
$160 < h \leqslant 164$	1

By using the midpoints of each group, calculate an estimate for the mean height of the students.

(3 marks)

AQA 2007

Objectives

Examiners would normally expect students who get these grades to be able to:

D

construct a histogram (frequency diagram) with equal class intervals

construct and interpret an ordered stem-and-leaf diagram

construct and interpret line graphs

C

construct a frequency polygon

B

construct and interpret a cumulative frequency diagram for continuous or grouped data

use a cumulative frequency diagram to estimate median and inter-quartile range

construct and interpret a box plot

compare two sets of data using a box plot referencing average and spread

A

construct a histogram with unequal class intervals

A*

interpret a histogram with unequal class intervals.

Key terms

stem-and-leaf diagram
back-to-back stem-and-leaf diagram
line graph
frequency polygon
frequency diagram
histogram

cumulative frequency diagram
cumulative frequency
lower quartile
upper quartile
inter-quartile range

Selling more phones than ever

May 2010
48 sold

June 2010
96 sold

Did you know?

Misleading diagrams

Charts and diagrams can sometimes be misleading.

Look at the diagram showing mobile phone sales in May and June. The sales in June are twice that in May and the dimensions of the phone in June are twice that in May – but does the advert look correct?

You need to be very careful to make sure your work is not misleading.

You should already know:

✔ the measures of average: mean, mode and median

✔ how to find and work with the range

✔ types of data, e.g. discrete and continuous, qualitative and quantitative

✔ the meaning of inequality signs

✔ how to work with percentages and fractions.

Learn... 7.1 Stem-and-leaf diagrams

Stem-and-leaf diagrams are a useful way of representing data.

They are used to show discrete data, or continuous data that has been rounded.

Stem-and-leaf diagrams need a key to show the 'stem' and 'leaf'.

For two-digit numbers the first digit is the stem and the second digit is the leaf.

It is often useful to provide an ordered stem-and-leaf diagram where the items are placed in order.

```
1 | 1  6  7  8  9
2 | 2  2  7  7  7  8  9
3 | 1  4  6
```

Key: 3 | 1 represents 31

If the numbers are decimals such as 5.4, the stem would be the 5 and the leaves would represent the decimal parts.

Example: A sample of 25 children in a primary school record how many portions of fruit and vegetables they eat in a week.

23 37 14 32 42 38 15 33 27 20 31 19 18

26 25 38 31 32 28 34 25 22 17 12 22

a   Draw an ordered stem-and-leaf diagram for the data.

b   Work out the range of the results.

c   Work out the median number of portions eaten.

Solution: **a**   The data runs from 12 to 42 and the values are all tens and units.

The stem will represent the tens and the leaf will represent the units.

The unordered stem-and-leaf diagram looks like this.

Unordered stem-and-leaf diagram showing portions of fruit and vegetables eaten

```
1 | 4  5  9  8  7  2
2 | 3  7  0  6  5  8  5  2  2
3 | 7  2  8  3  1  8  1  2  4
4 | 2
```

Key: 4 | 2 represents 42 portions of fruit and vegetables

The ordered stem-and-leaf diagram looks like this:

Ordered stem-and-leaf diagram showing portions of fruit and vegetables eaten

```
1 | 2  4  5  7  8  9
2 | 0  2  2  3  5  5  6  7  8
3 | 1  1  2  2  3  4  7  8  8
4 | 2
```

Key: 4 | 2 represents 42 portions of fruit and vegetables

b   Range = highest value − lowest value

= 42 − 12

= 30

c   The median is the middle value when values are put in order.

In an ordered stem-and-leaf diagram the data are already ordered.

There are 25 values so the median is the $\left(\frac{25 + 1}{2}\right)$th value or the 13th value.

Counting along this gives 26 portions. (Note that if the total was much above 25 you can simply halve the frequency to find the median, as when the sample size is large the difference is negligible.)

AQA *Examiner's tip*

Don't forget to include the stem number in your answer.
For example, here you must say that the median is 26, not 6.

Example: Two data sets can be shown at the same time on a **back-to-back stem-and-leaf diagram**.

This example of a back-to-back stem-and-leaf diagram compares the portions of fruit eaten by boys and girls. (Each side of the diagram then needs a label.)

	Boys					Girls			
	9	4	2	1	5	7	8		
5	3	2	0	2	2	5	6	7	8
8	4	2	2	1	3	1	3	7	8
				4	2				

Key: 9 | 1 represents 19   Key: 1 | 7 represents 17

Notice the leaves run backwards in order of size on the left.

This also means you need a key for **each side** of the diagram.

Practise...   **7.1 Stem-and-leaf diagrams**

D

1   The prices paid for a selection of items from a supermarket are as follows.

45p   32p   38p   21p   66p   54p   60p   44p   35p   42p   44p

a   Show the data in an ordered stem-and-leaf diagram.

b   What was the range of the prices paid?

c   What was the median price paid?

d   What was the modal price paid?

e   What percentage of the prices were under 50p?

2   The marks obtained by some students in a test were recorded as follows.

8   20   9   21   18   22   19   13   22   24
14   9   25   10   19   20   17   14   12

a   Show this information in an ordered stem-and-leaf diagram.

b   What was the highest mark in the test?

c   Write down the median of the marks in the test.

d   Write down the range of the marks in the test.

e   The pass mark for the test was 15 marks.
What fraction of the students passed the test?

3 The times taken to complete an exam paper were recorded as follows.

2 h 12 min   1 h 53 min   1 h 26 min   2 h 26 min   1 h 50 min
1 h 46 min   2 h 05 min   1 h 43 min   1 h 49 min   2 h 10 min
1 h 49 min   1 h 55 min   2 h 06 min   1 h 57 min

a Convert all the times to numbers.

b Show the converted data in an ordered stem-and-leaf diagram.

4 Look again at the data in Question 3.

Convert all the times to minutes.

Draw a new ordered stem-and-leaf diagram for this new data.

Comment on the similarities or otherwise of the new diagram with the one from Question 3.

! 5 The number of visitors to a small museum each day in July was counted.

The set of data has the following.

Minimum value of 23   Maximum value of 65
Median of 44   Modes of 42 and 55

Draw a possible stem-and-leaf diagram, making up data values which satisfy these conditions.

? 6 A village football team played 32 games during one season.

The numbers of spectators for the first 31 games are shown in the stem-and-leaf diagram.

```
18 | 4  6
19 | 0  2  3  5
20 | 3  3  7  9  9  9  9
21 | 0  1  5  5  6  7  8  8  9
22 | 3  4  4  6  9
23 | 2  5
24 | 7  8
```

Key: 18 | 4 represents 184 spectators

a The number of spectators at the 32nd game increases the range by 14.
Work out two possible values for the number of spectators at the 32nd game.

b Do either of these possible values affect the median of the number of spectators after 31 games? Explain your answer.

c Do either of the possible values found in part **a** affect the modal number of spectators after 31 games? Explain your answer.

⚙ 7 Emma is investigating this hypothesis:
'Girls take longer to complete an exercise than boys'.
She collects the data shown in this back-to-back stem-and-leaf diagram.

```
           Girls       |   |       Boys
    7 7 6 5 4 2 2  |  1  |  1 6 7 8 9
        7 6 4 3 2 1  |  2  |  2 2 7 7 7 8 9
              7 0  |  3  |  1 4 6
```

Key: 3 | 2 represents 23 minutes   Key: 3 | 4 represents 34 minutes

Compare the time taken by girls and boys to complete the exercise.

Write some conclusions that Emma might make about her hypothesis.

8　Declan is investigating reaction times for Year 7 and Year 11 students.

In an experiment he obtains these results.

Year	Times (tenths of a second)														
7	18	19	09	28	10	04	11	14	15	18	09	27	28	06	05
11	07	20	09	12	21	17	11	12	15	08	09	12	08	16	19

a　Show this information in a back-to-back stem-and-leaf diagram.

b　Declan thinks Year 7 have quicker reaction times than Year 11.
Use your diagram to show whether he is correct.

c　Explain how Declan might improve his experiment.

Learn... 7.2 Line graphs and frequency polygons

Line graphs

A **line graph** is a series of points joined with straight lines.

Line graphs show how data change over a period of time.

Here is an example of a line graph showing shop sales over a period of time.

Frequency polygons

A **frequency polygon** is a way of showing continuous grouped data in a diagram.

Points are plotted at the midpoint of each class interval.

For a frequency polygon, the groups may have equal or unequal widths.

The frequency polygon is an example of a **frequency diagram**.

Another type of frequency diagram is a **histogram**.

Histograms are discussed in Learn 7.5

Example:　50 people were asked how long they had to wait for a train.

The table below shows the results.

Time, t (minutes)	Frequency
$5 \leqslant t < 10$	16
$10 \leqslant t < 15$	22
$15 \leqslant t < 20$	11
$20 \leqslant t < 25$	1

Draw a frequency polygon to represent the data.

Solution: For a frequency polygon, the points are plotted at the midpoint of each class interval.

E.g. the midpoint for $5 \leqslant t < 10$ is $\dfrac{5 + 10}{2} = 7.5$

> **AQA** *Examiner's tip*
>
> Make sure that the axes are labelled with a continuous scale and not the class intervals.

Frequency polygon to show waiting times

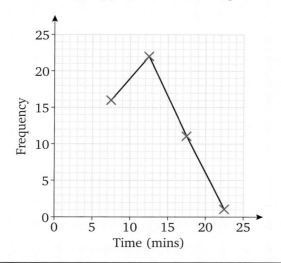

> **AQA** *Examiner's tip*
>
> There is no need to draw lines beyond the first and last plots.

7.2 Line graphs and frequency polygons

Practise...

D C B A A*

1   A new freezer is switched on for the first time at 9 o'clock one morning.

The temperature is noted every 10 minutes.

The readings (°C) for the next three hours are:

17.4	15.1	12.3	9.2	6.6	3.0	0.2	−2.3	−4.4
−7.1	−9.9	−12.0	−14.5	−16.6	−18.3	−19.1	−19.5	−19.6

a   Draw a fully labelled line graph to show these data.

b   What do you think the temperature of the freezer will be 24 hours later? Give a reason for your answer.

2   The table shows the times of runners in a fun run.

Time, t (minutes)	Frequency
5 up to 10	40
10 up to 15	125
15 up to 20	100
20 up to 25	55
25 up to 30	15

a   How many runners took part in the fun run?

b   Draw a frequency polygon to represent the data.

D

C

C

3

a The table shows the time taken for 100 students to solve a simple puzzle.

Times for 100 students	
Time, x (seconds)	Frequency
$10 \leqslant x < 20$	30
$20 \leqslant x < 30$	35
$30 \leqslant x < 40$	20
$40 \leqslant x < 50$	10
$50 \leqslant x < 60$	5

Draw and label the frequency polygon for these students.

b This table shows the time taken for 100 adults to solve the same puzzle.

Times for 100 adults	
Time, x (seconds)	Frequency
$10 \leqslant x < 20$	40
$20 \leqslant x < 30$	22
$30 \leqslant x < 40$	15
$40 \leqslant x < 50$	13
$50 \leqslant x < 60$	10

On the same axes draw and label the frequency polygon for these adults.

c Write down two comparisons shown by the frequency polygons between the students and the adults.

4 The frequency polygons show the times taken by sprinters in a series of 200 m races at a school.

The data is shown for boys and girls separately.

Compare the times of the boys and girls.

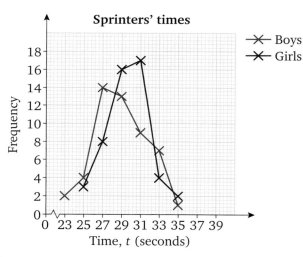

5 The principal of a college thinks that attendance becomes worse as the week progresses.

The table shows the number of students present during morning and afternoon registration.

Day	Mon	Mon	Tue	Tue	Wed	Wed	Thu	Thu	Fri	Fri
Session	am	pm	am	pm	am	pm	am	pm	am	pm
Number	220	210	243	215	254	218	251	201	185	152

a Show this information on a graph.

b There are 260 students in the college.
Work out the percentage attending each registration session.
Give your answers to the nearest whole number.

c Do your answers to parts **a** and **b** support the principal?
Explain your answer.

! 6   The duration of the 25 most popular films of 2008 are as follows.

1 hour 54 minutes	2 hours 3 minutes	1 hour 34 minutes	2 hours 23 minutes
2 hours 22 minutes	2 hours 12 minutes	1 hour 23 minutes	1 hour 49 minutes
1 hour 44 minutes	2 hours 43 minutes	2 hours 1 minute	2 hours 10 minutes
1 hour 30 minutes	1 hour 39 minutes	1 hour 28 minutes	2 hours 54 minutes
1 hour 39 minutes	1 hour 57 minutes	2 hour 2 minutes	1 hour 21 minutes
1 hour 40 minutes	3 hours 6 minutes	2 hours 29 minutes	1 hour 52 minutes
2 hours 9 minutes			

Use suitable class groupings to draw a frequency diagram for these data.

Learn... 7.3 Cumulative frequency diagrams

A **cumulative frequency diagram** (or cumulative frequency curve) is used to estimate the median and quartiles of a set of data.

To find the **cumulative frequency**, you add the frequencies in turn to give you a 'running total'.

Cumulative frequencies are plotted at the upper class bound. The upper class bound is the highest possible value for each class interval.

The cumulative frequency diagram is formed by joining the points with a series of straight lines or a smooth curve.

The total cumulative frequency can be divided by four to find the quartiles and the median. This is shown in the example that follows.

Example: The frequency distribution shows the length of 80 phone calls to an internet help line.

Time, t (minutes)	Frequency
$0 < t \leqslant 10$	9
$10 < t \leqslant 20$	23
$20 < t \leqslant 30$	31
$30 < t \leqslant 40$	12
$40 < t \leqslant 50$	5

a   Construct a column of cumulative frequencies.

b   Draw a cumulative frequency diagram for the data.

c   Use your diagram to estimate:

   i   the median length of phone call

   ii   the **lower quartile**

   iii   the **upper quartile**

   iv   the **inter-quartile range**.

d   Estimate the percentage of calls over 25 minutes.

Solution: **a**   It is useful to add an extra column to the table.
         This can be used to show the cumulative frequencies.

Time, t (minutes)	Frequency	Cumulative frequency
$0 < t \leqslant 10$	9	9
$10 < t \leqslant 20$	23	$9 + 23 = 32$
$20 < t \leqslant 30$	31	$32 + 31 = 63$
$30 < t \leqslant 40$	12	$63 + 12 = 75$
$40 < t \leqslant 50$	5	$75 + 5 = 80$

AQA *Examiner's tip*

Check your final total by adding up all the frequencies.

$9 + 23 + 31 + 12 + 5 = 80$ so the total cumulative frequency is correct.

(Notice the value 80 was given in the question.)

b The upper class bounds for each interval are shown below.

Time, t (minutes)	Upper class bound
$0 < t \leqslant 10$	10
$10 < t \leqslant 20$	20
$20 < t \leqslant 30$	30
$30 < t \leqslant 40$	40
$40 < t \leqslant 50$	50

> AQA **Examiner's tip**
>
> To draw a cumulative frequency diagram, cumulative frequencies are plotted at the upper class bound for each class interval.

The points to be plotted are (10, 9), (20, 32), (30, 63), (40, 75) and (50, 80).

(0, 0) can also be plotted as no calls are below zero minutes long.

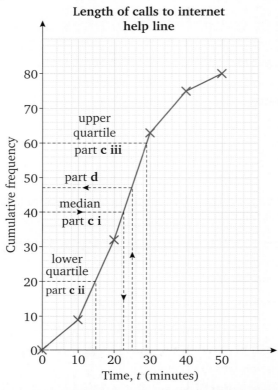

Length of calls to internet help line

c **i** The median is read off at the halfway point in the whole of the data set.
There are 80 values, so the median is the $\frac{1}{2} \times$ 80th value = 40th value
From the graph, median = 22.5 minutes
(Note the median is the $\frac{(n + 1)}{2}$th value which is $\frac{81}{2} = 40.5$ but when n is large it is simpler to just halve the cumulative frequency. The difference in outcome is negligible. This idea is also used in the position of the quartiles.)

ii The lower quartile is read off at the point one quarter along the data.
There are 80 values, so the lower quartile is the $\frac{1}{4} \times$ 80th value = 20th value
From the graph, lower quartile = 15 minutes

iii The upper quartile is read off at the point three quarters along the data set.
There are 80 values, so the upper quartile is the $\frac{3}{4} \times$ 80th = 60th value
From the graph, upper quartile = 29 minutes

iv The inter-quartile range is the difference between the upper and lower quartiles. It gives you the spread of the middle 50% of the data and is important for comparing the spread of different data sets.
The inter-quartile range = upper quartile − lower quartile
$$= 29 - 15$$
$$= 14 \text{ minutes}$$
(This means there is a 14 minute range across the middle 50% of the data.)

d   To find an estimate of the percentage of calls over 25 minutes:

Draw a line from 25 minutes on the horizontal axis to the cumulative frequency graph.

From the graph, reading across = 47 calls.

There are 47 calls **below** 25 minutes.

There are 80 − 47 = 33 calls **above** 25 minutes.

This represents $\frac{33}{80}$ calls = $\frac{33}{80} \times 100\% = 41.25\%$

Practise...   7.3 Cumulative frequency diagrams

B

1   Complete the cumulative frequency columns for these frequency tables.

a

Height, h (cm)	Frequency	Cumulative frequency
$100 < h \leqslant 120$	5	5
$120 < h \leqslant 140$	12	5 + 12 = 17
$140 < h \leqslant 160$	10	17 + 10 =
$160 < h \leqslant 180$	7	
$180 < h \leqslant 200$	4	

b

Weight, w (kg)	Frequency	Cumulative frequency
$10 < w \leqslant 11$	300	300
$11 < w \leqslant 12$	254	554
$12 < w \leqslant 13$	401	
$13 < w \leqslant 14$	308	
$14 < w \leqslant 15$	126	

c

Time, t (seconds)	Frequency	Cumulative frequency
$10 \leqslant t < 30$	43	43
$30 \leqslant t < 50$	65	
$50 \leqslant t < 70$	72	
$70 \leqslant t < 90$	55	

d

Height, h (feet)	Frequency	Cumulative frequency
$100 \leqslant h < 150$	1	
$150 \leqslant h < 200$	15	
$200 \leqslant h < 250$	34	
$250 \leqslant h < 300$	46	
$300 \leqslant h < 350$	16	
$350 \leqslant h < 400$	9	

2   For each of the tables in Question 1, draw a cumulative frequency diagram.

B

3   The cumulative frequency diagram shows the mass of 100 gerbils in a pet shop.

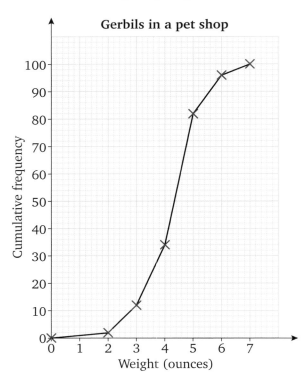

Gerbils in a pet shop

Use the diagram to estimate:

a   the median

b   the lower quartile

c   the upper quartile

d   the inter-quartile range

e   the percentage of gerbils under 2 ounces

f   the percentage of gerbils over 5.5 ounces.

4   The frequency distribution shows the lifetimes of 1000 light bulbs in hours.

Lifetime, l (hours)	Frequency	Cumulative frequency
$50 < l \leq 100$	80	
$100 < l \leq 150$	240	
$150 < l \leq 200$	390	
$200 < l \leq 250$	200	
$250 < l \leq 300$	90	

a   Copy and complete the table.

b   Draw the cumulative frequency diagram.

c   Use your diagram to estimate:

     i   the median

     ii   the lower quartile

     iii   the upper quartile

     iv   the inter-quartile range

     v   Estimate the percentage of light bulbs lasting beyond one week of continuous use.

d   Explain why your answers to the parts in **c** are estimates.

! 5   The data represent the weights of 100 newborn babies.

Weight, w (kg)	Frequency	Cumulative frequency
$0 < w \leqslant 1$	1	1
$1 < w \leqslant 2$	a	18
$2 < w \leqslant 3$	72	b
$3 < w \leqslant 4$	c	98
$4 < w \leqslant 5$	d	e

a   Find the values indicated with letters in the table.

b   Draw a cumulative frequency diagram for the data.

c   Estimate the inter-quartile range of the babies' weights.

6   Speeds on some roads are constantly monitored. If the upper quartile of actual car speeds is above the speed limit, speed cameras are considered to enforce the speed limit.

a   For these data, speed cameras were not considered.
     What was the speed limit?

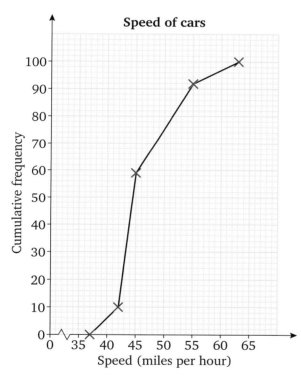

b   A different road has a speed limit of 70 miles per hour.

     Comment on the need for consideration of speed cameras in light of the data in the table.

     Give a reason for your answer.

Speed, s (mph)	Frequency
$55 < s \leqslant 60$	27
$60 < s \leqslant 65$	36
$65 < s \leqslant 72$	40
$72 < s \leqslant 80$	11
$80 < s \leqslant 95$	6

7   The hourly pay of employees at a certain company is shown in the table.

The company is situated in a country outside of the UK.

The minimum wage is $\frac{2}{3}$ of the median pay.

a   Estimate the minimum wage of this country.

b   Are there any employees working for the minimum wage? Justify your answer and if so, estimate the percentage doing so.

Hourly pay, p (euro)	Frequency
$4 < p \leqslant 5$	17
$5 < p \leqslant 6$	56
$6 < p \leqslant 7$	55
$7 < p \leqslant 8$	101
$8 < p \leqslant 9$	39

8 The cumulative frequency diagram represents the length of service of workers at a company.

Use the diagram to complete the missing information in this company report.

The company has _____ employees and all but six have been with the company for at least _____ years.

The median length of service is _____ years. This is an increase of 2.6 years on the figure of two years ago which was _____ years.

The inter-quartile range is 10% less than two years ago when the figure was _____ years.

I am pleased to be able to recommend _____ workers for long service awards as they have worked for more than 35 years.

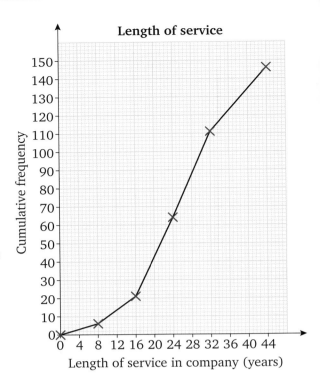

Learn... 7.4 Box plots

A box plot (sometimes called a box-and-whisker diagram) is another way to show information about a frequency distribution.

The box plot provides a visual summary of information.

It can be used to compare two or more distributions.

The box plot shows the following information.

* the minimum and maximum values
* the lower and upper quartiles
* the median

Example: The box plot shows the length of songs in a record collection (recorded in minutes).

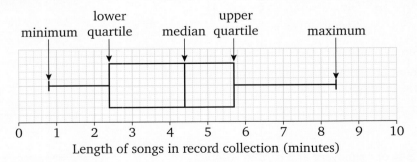

Write down:

a the minimum song length

b the maximum song length

c the median song length

d the lower quartile song length

e the upper quartile song length.

Solution: From the diagram, you can see that:

a   the minimum song length was 0.8 minutes long

b   the maximum song length was 8.4 minutes long

c   the median song length was 4.4 minutes long

d   the lower quartile song length was 2.4 minutes long

e   the upper quartile song length was 5.7 minutes long.

Box plots are sometimes drawn at the base of a cumulative frequency diagram.

This is because the measures are often estimated from the diagram.

Box plots can be put together to compare two or more data sets.

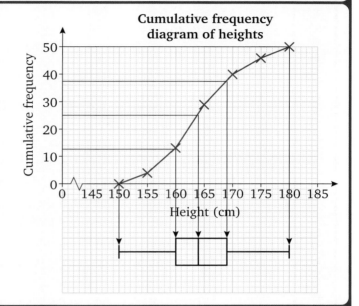

Cumulative frequency diagram of heights

Example: A group of men competed against a group of women in a series of puzzles.

The times to finish obtained by each person are summarised in the table.

Time, t (minutes)	Male frequency	Female frequency
$0 < t \leqslant 10$	3	5
$10 < t \leqslant 20$	11	16
$20 < t \leqslant 30$	35	21
$30 < t \leqslant 40$	27	29
$40 < t \leqslant 50$	4	9

a   Use a suitable diagram to obtain estimates of measures for a box plot.

b   Draw the box plot and compare the performances of men and women.

> AQA   *Examiner's tip*
>
> You will see the words 'use a suitable diagram' in examination questions. Here a cumulative frequency diagram is the suitable diagram but you will have to choose.

Solution: **a**   A cumulative frequency diagram is a suitable diagram to obtain estimates of measures for a box plot.

Time, t (minutes)	Male frequency	Female frequency	Male cumulative frequency	Female cumulative frequency
$0 < t \leqslant 10$	3	5	3	5
$10 < t \leqslant 20$	11	16	14	21
$20 < t \leqslant 30$	35	21	49	42
$30 < t \leqslant 40$	27	29	76	71
$40 < t \leqslant 50$	4	9	80	80

The upper class values will be 10, 20, 30, …, and so on.

Drawing both graphs on the same axes:

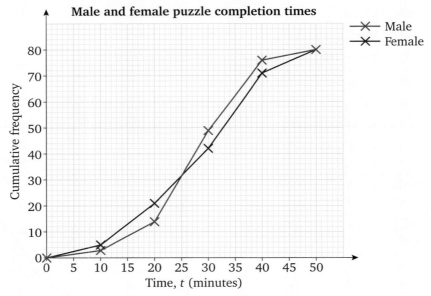

Male and female puzzle completion times

─✕─ Male
─✕─ Female

b To obtain the box plots, use the cumulative frequency diagrams.

The median is read off at the halfway point in the whole of the data set.

There are 80 values, so the median is the $\frac{1}{2} \times$ 80th value = 40th value

The lower quartile is read off at the halfway point in the bottom half of the data.

There are 80 values, so the lower quartile is the $\frac{1}{4} \times$ 80th value = 20th value

The upper quartile is read off at the halfway point in the top half of the data set.

There are 80 values, so the upper quartile is the $\frac{3}{4} \times$ 80th = 60th value

The minimum for both groups has to be taken as 0.

The maximum for both groups has to be taken as 50

Putting all the results into a table you get estimates as follows:

	Minimum	Lower quartile	Median	Upper quartile	Maximum
Data for women	0	19.5	29	36	50
Data for men	0	22	27.5	34	50

Showing this information as a pair of box plots:

Box plot showing average times to complete puzzles

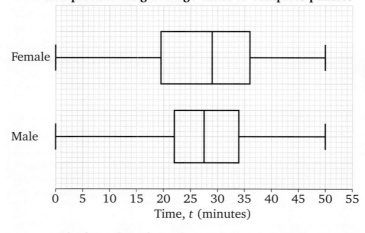

The box plots show that on average men do better.

The median shows that women took a higher average time.

The box plots also show that men are more consistent than women, as the inter-quartile range (the width of the box part) is narrower.

Also women have more really quick times and more really slow times.

Practise... 7.4 Box plots

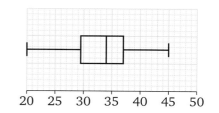

1 **a** From this box plot, write down the following values.

  i minimum

  ii lower quartile

  iii median

  iv upper quartile

  v maximum

 b Calculate the inter-quartile range.

2 A set of data has these measures.

	Minimum	Lower quartile	Median	Upper quartile	Maximum
Data value	10	29	37	50	66

 a Draw a box plot for these data.

 b Explain why the inter-quartile range is 21.

3 The box plots below show the ages of all of the people in two villages.

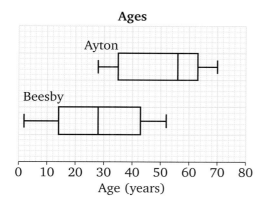

 a Complete the table for the two villages.

	Ayton	Beesby
Minimum age		
Maximum age		
Lower quartile age		
Upper quartile age		
Median		

 b Compare the ages in the two villages.

4 The frequency distribution shows the times taken by runners in the 'Race for Life'.

 a Draw a cumulative frequency diagram for these data.

 b Draw a box plot for these data.

 c The previous year, the median race time had been 37.1 minutes with an inter-quartile range of 9.9 minutes. Compare the results for this year and the previous year.

Time, t (minutes)	Frequency
25 up to 30	27
30 up to 35	215
35 up to 40	307
40 up to 45	147
45 up to 50	104

B

5 The cumulative frequency diagram opposite shows the waiting times at a main post office.

The box plot below shows the waiting times at a village post office.

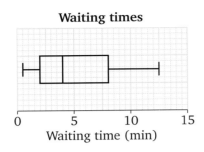

Waiting times

0   5   10   15
Waiting time (min)

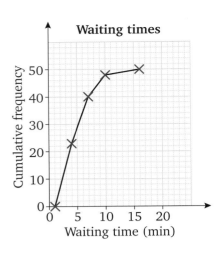

Waiting times

Compare the waiting time at the two post offices.

! 6 Match the cumulative frequency distribution with the box plot.

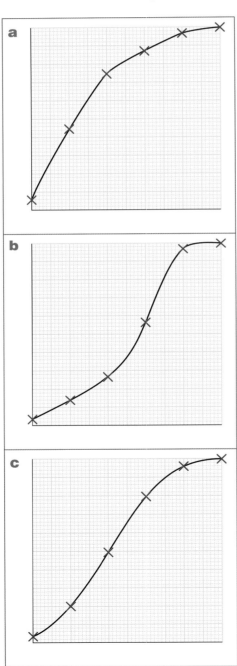

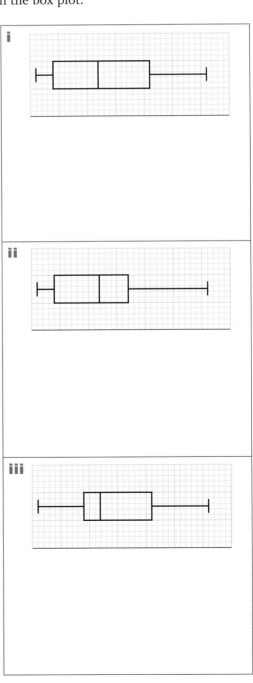

7   The box plot shows information about hourly sales in a small corner shop over a period of time.

Hourly sales

Hourly sales (£)

The shop is taken over by a new manager.

One month later, the following data are collected.

Hourly sales (£)	Frequency
£5 up to £10	11
£10 up to £15	23
£15 up to £20	63
£20 up to £25	2
£30 up to £35	1

Write a report on the sales figures before and after the shop is taken over by the new manager. Include measures obtained from suitable diagrams.

Learn...   7.5 Histograms

Histograms

A histogram is a way of showing continuous grouped data in a diagram.

The area of the bar represents the frequency.

For a histogram, the groups may have equal or unequal widths.

The histogram is an example of a frequency diagram.

Another type of frequency diagram is a frequency polygon.

Frequency polygons were discussed earlier in Learn 7.2 Line graphs and frequency polygons.

Histograms with equal group widths

In a histogram, the area of the bars represents the frequency.

If the group widths are equal, bars are drawn to the height of the frequency.

Example: 50 people were asked how long they had to wait for a train.

The table below shows the results.

Time, t (minutes)	Frequency
$5 \leqslant t < 10$	16
$10 \leqslant t < 15$	22
$15 \leqslant t < 20$	11
$20 \leqslant t < 25$	1

Draw a histogram to represent the data.

Solution: Use a continuous scale for the *x*-axis.

As the groups are all equal width, bars are drawn to the height of the frequency.

The areas will then be proportional to the frequency.

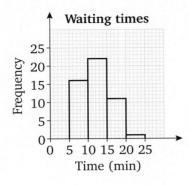

Waiting times

Histograms with unequal group widths

For a histogram, the groups may have equal or unequal widths.

In a histogram, the area of the bars represents the frequency.

If the group widths are unequal, bars are drawn to the height of the frequency density.

$$\text{Frequency density} = \frac{\text{frequency}}{\text{class width}}$$

Example: The table shows the waiting times in seconds for a phone to be answered in a call centre.

a   Draw a fully labelled histogram for these data.

b   Estimate the proportion of calls for which the waiting time is more than 4 seconds.

Time, t (seconds)	Frequency
$0 < t \leqslant 2$	38
$2 < t \leqslant 3$	32
$3 < t \leqslant 3.5$	30
$3.5 < t \leqslant 5$	45
$5 < t \leqslant 10$	55

Solution: **a**   Use a continuous scale for the *x*-axis.

As the groups are unequal width, bars are drawn to the height of the frequency density.

$$\text{Frequency density} = \frac{\text{frequency}}{\text{class width}}$$

It is useful to add extra columns to the table.

A column can be used to show the class width. This is the upper class bound take away the lower class bound.

The other column can be used to show the frequency density.

Time, t (seconds)	Frequency	Class width	Frequency density $= \dfrac{\text{frequency}}{\text{class width}}$
$0 < t \leqslant 2$	38	2	$38 \div 2 = 19$
$2 < t \leqslant 3$	32	1	$32 \div 1 = 32$
$3 < t \leqslant 3.5$	30	0.5	$30 \div 0.5 = 60$
$3.5 < t \leqslant 5$	45	1.5	$45 \div 1.5 = 30$
$5 < t \leqslant 10$	55	5	$55 \div 5 = 11$

Now the first bar is drawn from 0 to 2 to a height of 19 and so on.

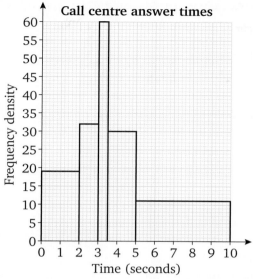

Call centre answer times

b   Above 4 seconds is all of the calls for the $5 < t \leqslant 10$ class and two thirds of the calls for the $3.5 < t \leqslant 5$ class (as 4 is one third along the length of the class).

This estimate is therefore $55 + \frac{2}{3} \times 45 = 55 + 30 = 85$ calls

Practise...   7.5 Histograms

1   The heights of people auditioning for a play are given in the table.

Height, h (cm)	Frequency
$100 < h \leqslant 120$	24
$120 < h \leqslant 140$	20
$140 < h \leqslant 160$	13
$160 < h \leqslant 180$	6
$180 < h \leqslant 200$	7

a   Draw a fully labelled histogram for these data.

b   What would be different about the histogram if the final class was $180 < h \leqslant 220$ instead of $180 < h \leqslant 200$?

2   The table shows the wages of workers in a factory.

Wages, x (£)	Frequency
$100 < x \leqslant 200$	120
$200 < x \leqslant 250$	165
$250 < x \leqslant 300$	182
$300 < x \leqslant 350$	197
$350 < x \leqslant 400$	40
$400 < x \leqslant 600$	6

a   Draw a suitable diagram for these data.

b   Estimate the number of workers earning more than £387.50.

A

3 Draw fully labelled histograms for these sets of data.

a

Time, t (hours)	Frequency
$0 < t \leqslant 20$	3
$20 < t \leqslant 30$	11
$30 < t \leqslant 35$	35
$35 < t \leqslant 45$	27
$45 < t \leqslant 95$	4

b

Speed, s (mph)	Frequency
$55 < s \leqslant 60$	25
$60 < s \leqslant 65$	35
$65 < s \leqslant 72$	42
$72 < s \leqslant 80$	12
$80 < s \leqslant 95$	6

c From each histogram estimate the median of the distribution.

> **Hint**
> The median is halfway along the data and so will divide the total area under the bars into two equal parts.

4 The distribution of ages of people at a hotel is shown in the histogram below.

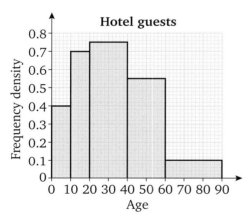

Complete the following table for the distribution.

Age, a (years)	$0 < a \leqslant 10$	$10 < a \leqslant 20$	$20 < a \leqslant 40$	$40 < a \leqslant 60$	$60 < a \leqslant 90$
Number	4				

5 A ski resort regularly measures the depth of snow.

The frequency table shows the depth of snow throughout the season.

a Criticise the labelling of the table.

b Construct a fully labelled histogram for the data.

Depth, d (cm)	Frequency
10–50	12
50–80	27
80–100	28
100–120	19
120–200	64

6 The table shows some of the percentages of a sample of tins of beans within various groups of weights.

a Copy and complete the table.

b Draw a fully labelled histogram to show all the data appropriately.

Weight, w (g)	Percentage
$390 < w \leqslant 400$	10
$400 < w \leqslant 401$	26
$401 < w \leqslant 402$	17
$402 < w \leqslant 405$	33
$405 < w \leqslant 412$	

7 A restaurant group advertise that they sell a particular drink in 500 ml servings.

Over a period of time trading standards officers sample these drinks.

The table shows the volumes of the sample of drinks.

a Draw a fully labelled histogram for these data.

b Estimate the proportion of drinks which were under the advertised volume.

c Comment on your answer to part **b** in the context of the question.

Volume, v (ml)	Frequency
$490 < v \leqslant 495$	8
$495 < v \leqslant 498$	18
$498 < v \leqslant 501$	21
$501 < v \leqslant 505$	44
$505 < v \leqslant 510$	15

8 Virgil likes to read his morning paper for half an hour before going to work.

He has to leave for work at 8.15am.

The table shows the arrival times for the newspaper over the last few months.

a Draw a fully labelled histogram to show the data.

b Estimate the proportion of times:

i Virgil has time to read the newspaper for half an hour in the morning.

ii Virgil has to leave for work before the newspaper arrives.

Arrival time (am)	Frequency
7.00–7.20	12
7.20–7.40	42
7.40–7.55	45
7.55–8.05	23
8.05–8.30	15

9 The histogram shows the distribution of weights of apples.

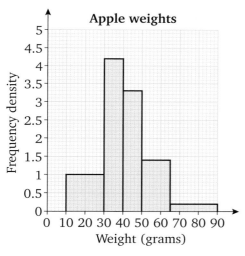

a Construct the frequency distribution for these weights.

b Calculate an estimate of the mean weight of apples.

Assess *k!*

1 Rod and Annette went fishing every month for six years.

They kept the following record of the number of fish caught each month.

One number needs to be filled in before you can draw a graph.

Construct a fully labelled histogram for the data.

Number caught	Frequency
$0 \leqslant n \leqslant 10$	3
$10 < n \leqslant 20$	17
$20 < n \leqslant 30$	31
$30 < n \leqslant 40$	19
$40 < n \leqslant 50$	?

D

D

2 The heights of 40 workers in a factory are given in the diagram below.

	Females			Males	
	9	14			
	9 8 2	15			
9 9 8 7 6 6 6 4 4 3 1	16	2 4 7 9 9			
8 7 5 3 3 2 2 1	17	2 2 4 5 5 8			
2	18	3 3 6 9			
	19	1			

Key: 7 | 16 represents 167 cm   Key: 18 | 3 represents 183 cm

a Explain why the median male height is 174.5 cm.

b Show that only about 19% of males are shorter than the median female height.

c Compare the ranges of the male and female heights.

d Produce a similar diagram for data from your class.

D
C

3 The number of workers in the canteen in a hospital is shown in the line graph.

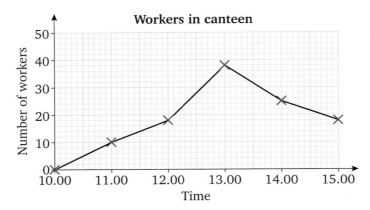

a At what time is the canteen busiest?

b What time do you think the canteen opens?
Give a reason for your answer.

c Complete this table:

Time	10.00	11.00	12.00	13.00	14.00	15.00
Number of workers in canteen						

d Is a line graph a suitable diagram for these data?
Give a reason for your answer.

e Display the data in a suitable diagram of your choice.

B

4 The ages of passengers on a train are shown in the frequency table.

a Calculate the cumulative frequencies for these data.

b Hence draw a cumulative frequency diagram.

c Use your diagram to estimate the following:

i median

ii lower quartile

iii upper quartile

iv inter-quartile range.

Age, a (years)	Frequency
$0 \leqslant a < 5$	11
$5 \leqslant a < 10$	32
$10 \leqslant a < 15$	28
$15 \leqslant a < 20$	35
$20 \leqslant a < 30$	45
$30 \leqslant a < 50$	58
$50 \leqslant a < 65$	115
$65 \leqslant a < 90$	76

5 The cumulative frequency diagram shows the ages of passengers on an aircraft.

Use this diagram and your results from Question 4 to draw two box plots on the same axis.

Compare the ages of travellers on the train and the aircraft.

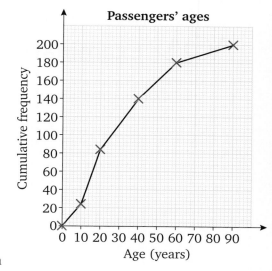

6 **a** Draw a fully labelled histogram using the train data from Question 4.

b Use the histogram to obtain an estimate of the median age. How does this compare to your answer in Question 4 part **c i**?

7 **a** The frequency distribution shows the heights of 150 trees in a country park.

Draw **two** different frequency diagrams for these data.

Country park	
Height, h (metres)	**Frequency**
$5 < h \leqslant 10$	48
$10 < h \leqslant 15$	55
$15 < h \leqslant 20$	27
$20 < h \leqslant 25$	16
$25 < h \leqslant 30$	4

b The frequency distribution shows the heights of 90 trees in a local park.

Donna wants to draw a diagram to compare the heights of the two sets of trees.

She draws the frequency polygon for the local park.

Local park	
Height, h (metres)	**Frequency**
$0 < h \leqslant 5$	7
$5 < h \leqslant 10$	51
$10 < h \leqslant 15$	23
$15 < h \leqslant 20$	3
$20 < h \leqslant 30$	6

Tree heights

What mistake has she made?

c Draw suitable diagrams to compare the heights of the two sets of trees.

d Nick wants to compare the medians and quartiles for the two sets of trees. Draw suitable diagrams to make estimates of the medians and quartiles.

e Use your information to compare the heights of the two sets of trees.

AQA Examination-style questions

1   **a**   This histogram shows the test scores of 100 female students.

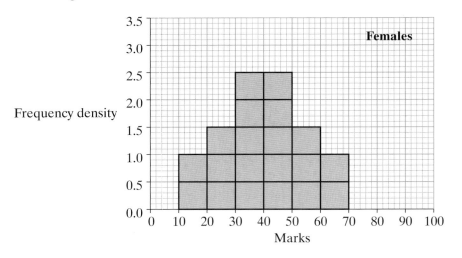

   i   What is the median score?   (1 mark)
   ii   What is the inter-quartile range?   (1 mark)

b   This histogram is incomplete.
It shows some of the test scores for 100 male students.
The median test score for males is the same as for females.
The upper quartile for the males is 50.

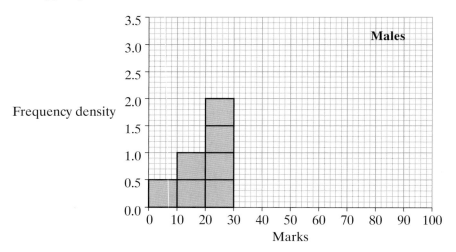

   i   What is the lower quartile for the male students?   (1 mark)
   ii   Complete a possible histogram.   (3 marks)

                                                                              AQA 2004

Scatter graphs

Objectives

Examiners would normally expect students who get these grades to be able to:

D

draw a scatter graph by plotting points on a graph

interpret the scatter graph

C

draw a line of best fit on the scatter graph

interpret the line of best fit

identify the type and strength of the correlation.

Did you know?

Scatter graphs

Scatter graphs are frequently used in medical research to test for relationships. For example, a study of office workers found that those with a stressful job had higher blood pressure. Scatter graphs can also be used to test the effects of drugs on lowering blood pressure.

Key terms

coordinate
scatter graph
correlation
positive correlation
negative correlation
zero or no correlation
outlier
line of best fit

You should already know:

✔ how to use **coordinates** to plot points on a graph

✔ how to draw graphs including labelling axes and adding a title.

Learn... 8.1 Interpreting scatter graphs

Scatter graphs (or scatter diagrams) are used to show the relationship between two sets of data.

Correlation measures the relationship between two sets of data.

It is measured in terms of **type** and **strength** of correlation.

Type of correlation

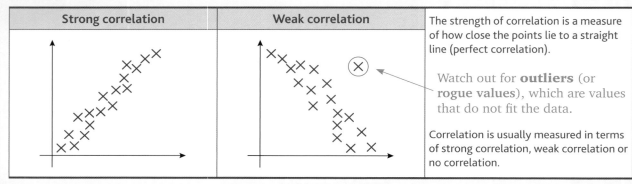

Positive correlation	Negative correlation	Zero or no correlation
Positive correlation As one set of data increases, the other set of data increases.	**Negative correlation** As one set of data increases, the other set of data decreases.	**Zero or no correlation** There is no obvious relationship between the two sets of data.

Example:

Temperature against ice cream sales. As the temperature increases, the number of ice cream sales increases.

Example:

Temperature against sales of coats. As the temperature increases, the sale of coats decreases.

Example:

Temperature against toothpaste sales. There is no obvious relationship between temperature and toothpaste sales.

Strength of correlation

Strong correlation	Weak correlation	

The strength of correlation is a measure of how close the points lie to a straight line (perfect correlation).

Watch out for **outliers** (or **rogue values**), which are values that do not fit the data.

Correlation is usually measured in terms of strong correlation, weak correlation or no correlation.

Example:

The graph shows the temperature and sales of ice cream.

Describe the relationship between the temperature and sales of ice cream.

> **Bump up your grade**
>
> You need to be able to describe the type and strength of the relationship between two sets of data to get a Grade C.

Solution:

You can see that there is a relationship between the temperature and sales of ice cream.

As the temperature increases, the sales of ice cream increase.

There is a **strong positive** correlation between the temperature and the sales of ice cream. This is probably because as the temperature goes up, people want to eat more ice cream.

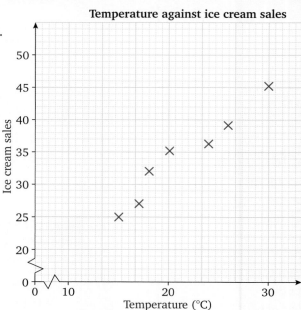

Temperature against ice cream sales

Practise... 8.1 Interpreting scatter graphs (k!)

D C B A A*

D

1 For each of the following:

a describe the type and strength of correlation

b write a sentence explaining the relationship between the two sets of data.

i The hours of sunshine and the sales of iced drinks

ii The number of cars on a road and the average speed

iii The distance travelled and the amount of petrol used

iv The cost of a house and the number of bedrooms

v The amount of sunshine and the sale of umbrellas.

2 For each of these scatter graphs:

a describe the type and strength of correlation

b write a sentence explaining the relationship between the two sets of data (for example, the higher the rainfall, the heavier the weight of apples).

i

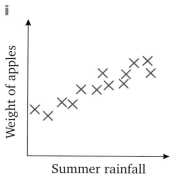

iii

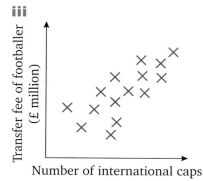

v

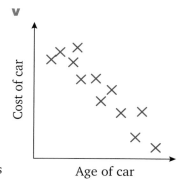

ii

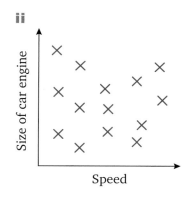

iv

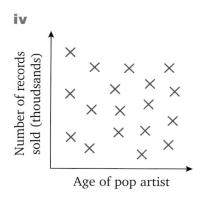

3 The table shows the ages and arm spans of seven students in a school.

Age (years)	16	13	13	10	18	10	15
Arm span (inches)	62	57	59	57	64	55	61

a Represent the data on a scatter graph.

b Describe the type and strength of correlation.

c Write a sentence explaining the relationship between the two sets of data.

D

4 The table shows the hours of sunshine and rainfall in 10 seaside towns.

Sunshine (hours)	Rainfall (mm)
650	11
400	30
530	28
640	11
520	24
550	20
480	26
600	15
550	16
525	23

a Represent the data on a scatter graph.

b Describe the type and strength of correlation.

c Write a sentence explaining the relationship between the two sets of data.

5 For each graph, write down two variables that might fit the relationship.

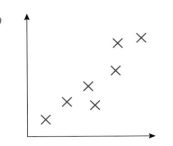 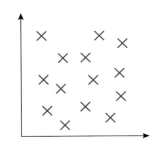

a   **b**   **c**

! 6 The scatter graph shows the ages and shoe sizes of a group of people.

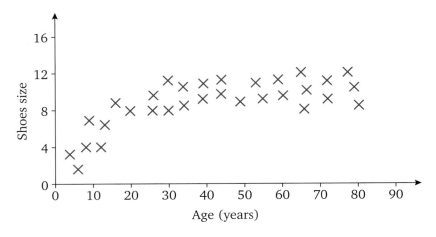

a Describe the type and strength of the correlation.

b Give a reason for your answer.

7 Ron is investigating the fat content and the calorie values of food at his local fast-food restaurant.

He collects the following information.

	Fat (g)	Calories
Hamburger	9	260
Cheeseburger	12	310
Chicken nuggets	24	420
Fish sandwich	18	400
Medium fries	16	350
Medium cola	0	210
Milkshake	26	1100
Breakfast	46	730

a   Describe the correlation between fat and calories.

b   Does the relationship hold for all the different foods? Give a reason for your answer.

> **Hint**
>
> If you are asked to describe correlation you should draw a scatter graph first, then describe the type and strength of correlation.

Learn...   8.2 Lines of best fit

A **line of best fit** is drawn to represent the relationship between two sets of data on a scatter graph.

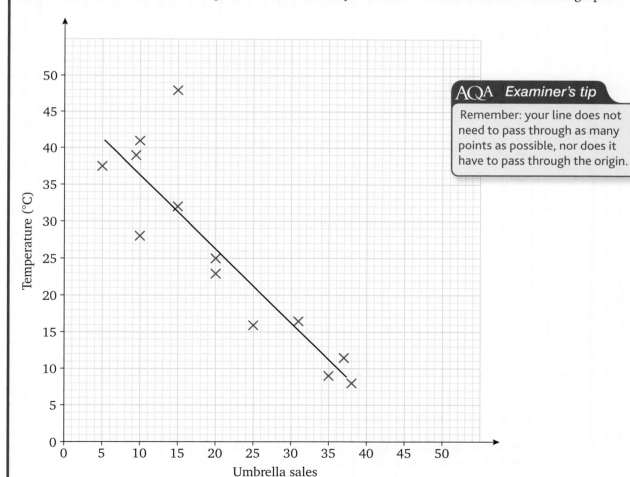

> AQA   *Examiner's tip*
>
> Remember: your line does not need to pass through as many points as possible, nor does it have to pass through the origin.

In this example, one of the values does not seem to fit the rest of the data. This is called an **outlier** or rogue value. Ignore these values when drawing a line of best fit.

You should draw the line of best fit so that:
- it gives a general trend for all of the data on the scatter graph
- it gives an idea of the strength and type of correlation
- there are roughly equal numbers of points above and below the line.

You can use the line of best fit to estimate missing data.

A line of best fit should only be drawn where the correlation is strong.

> **AQA Examiner's tip**
>
> The examiner will use a 'corridor of success' to check that your line of best fit is reasonable. This means that if your line is in a certain area on the graph, you will get the mark.

Example: The graph shows the number of hours revision and the number of GCSE passes for 10 students.

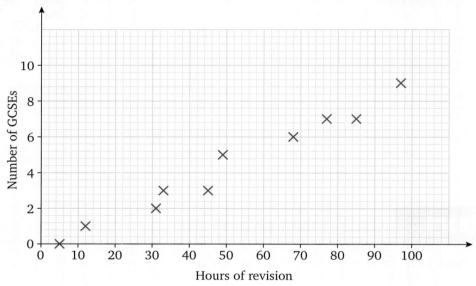

a Draw a line of best fit.

b Dinah studies for 60 hours. How many GCSE passes is she likely to get?

Solution: By drawing the line of best fit, you can use the graph to estimate the number of GCSE passes.

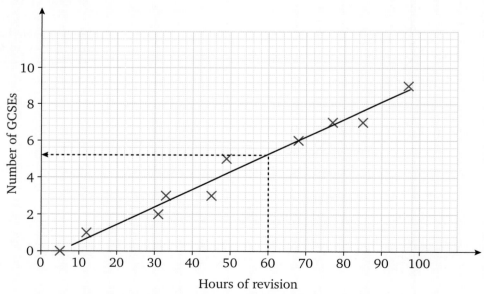

From the graph Dinah should expect to get five passes.

> **Bump up your grade**
>
> You will need to draw and interpret the line of best fit to get a Grade C.

Practise...   8.2 **Lines of best fit**   D C B A A*

D

1   The table shows the rainfall and the number of sunbeds sold in a day at a resort.

Amount of rainfall (mm)	0	1	2	5	6	9	11
Number of sunbeds sold	380	320	340	210	220	110	60

   a   Draw a scatter graph to represent this information.

   b   Draw a line of best fit and use it to estimate:
      i   the number of sunbeds sold when there is 4 mm of rainfall
      ii   the amount of rainfall if 100 sunbeds are sold.

2   The table shows the age and value of seven second-hand cars of the same model.

Age of car (years)	2	1	4	7	10	9	8
Value of car (£)	4200	4700	2800	1900	400	1100	2100

   a   Draw a scatter graph to represent this information.

   b   Draw a line of best fit and use it to estimate:
      i   the value of a car if it is 7.5 years old
      ii   the age of a car if its value is £3700.

3   Rob collects information on the temperature and the number of visitors to an art gallery.

Temperature (°C)	15	25	16	18	19	22	24	23	17	20	26	20
Number of visitors	720	180	160	620	510	400	310	670	720	530	180	420

   a   Draw a scatter graph to represent this information.

   b   Estimate:
      i   the number of people if the temperature is 24 °C
      ii   the temperature if 350 people visit the art gallery.

   c   Rob is sure that two sets of data are incorrect. Identify these two sets of data on your graph.

 4   The table shows the distances from the equator and average temperatures for 12 cities.

The distance is measured in degrees from the equator.

The temperature is measured in degrees Celsius.

   a   What do you notice?

   b   Dubai is 25° north of the equator. Use this to find the average temperature in Dubai.

   c   What other factors might affect temperatures?

City	Distance from equator (°)	Average temp. (°C)
Bangkok	13	28
Beijing	39	12
Boston	42	9
Cairo	30	22
Cape Town	33	17
Copenhagen	55	8
Gibraltar	36	19
Istanbul	40	14
London	51	10
Moscow	55	4
Mumbai	18	27
Perth	32	18

5 The graph shows the line for best fit for the relationship between house prices in 2006 and estimated house prices in 2020.

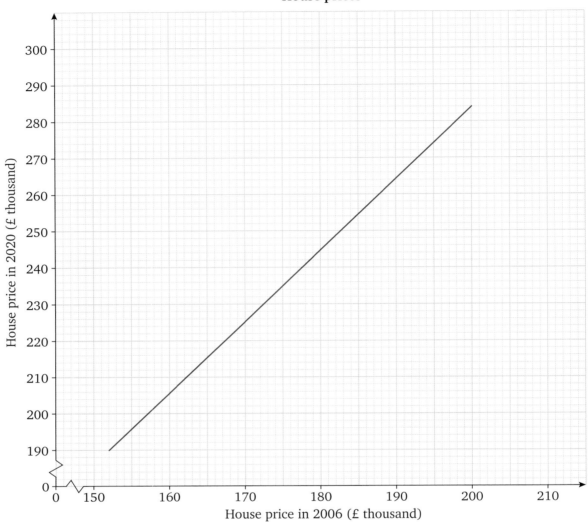

House prices

a Copy and complete the table, giving estimates for the missing values.

House price in 2006 (£ thousand)	170	175	190	200		
House price in 2020 (£ thousand)	225				305	355

b Rashid says his house price was £180 000 in 2006 and £270 000 in 2020. Is he correct? Give a reason for your answer.

c Find an estimate for the 2012 price of a house priced £155 000 in 2006.

d Find an estimate for the 2006 price of a house priced £280 000 in 2020.

e Which of these results is likely to be the most reliable? Give a reason for your answer.

6 Jenny collects information on the top speed and engine size of various motorbikes. Her results are shown in the table below.

Top speed (kph)	70	120	140	150	180	190	220	250	270	260	270	240
Engine size (cc)	50	250	350	270	400	440	600	800	950	900	1200	1000

a Draw a scatter graph of the results.

b   What do you notice about the correlation between speed and engine size?

c   Draw a curve of best fit and use this to estimate:

    i   the engine size if the top speed is 170 kph

    ii   the engine size if the top speed is 250 kph.

d   Which of these results is likely to be the most reliable?

    Give a reason for your answer.

8   Assess k!

1   The information below shows the marks of eight students in history and geography.

Student	A	B	C	D	E	F	G	H
History	25	35	28	30	36	44	15	21
Geography	27	40	29	32	41	48	17	20

Draw a scatter graph to represent this information and comment on the relationship between the history and geography marks.

2   The following table shows the hours of TV watched and test marks for 10 students.

Student	1	2	3	4	5	6	7	8	9	10
TV hours	4	7	9	10	13	14	15	20	21	25
Test mark	9	90	74	30	74	66	95	38	35	30

a   Draw a scatter graph to represent this information and comment on the relationship between the figures.

b   Two students do not seem to 'fit the trend'. Which ones are they? Explain why.

3   The table shows the relationship between the area (in thousands of km²) of some European countries and their populations (in millions).

	Monaco	Malta	Jersey	Netherl.	UK	Germ.	Italy	Switz.	Andorra	Denm.
Area	0.002	0.3	0.1	41	245	357	294	41	0.5	43
Population	0.03	0.4	0.09	16	61	82	59	7	0.08	5

	France	Austria	Turkey	Greece	Spain	Ireland	Latvia	Sweden	Norway	Iceland
Area	551	84	783	130	504	84	65	450	323	100
Population	62	8	74	11	45	6	2	9	5	0.3

Draw a scatter graph of these data and comment on the graph.

4 The table shows the distance jumped in long jump trials and the leg length of the jumpers.

Leg length (cm)	71	73	74	75	76	79	82
Distance jumped (m)	3.2	3.1	3.3	4.1	3.9	4	4.8

a Draw a scatter graph to represent this information.

b Use a line of best fit to estimate:

    i the leg length of an athlete who jumped a distance of 3.5 m

    ii the distance jumped by an athlete with a leg length of 85 cm.

c Explain why one of those estimates is more reliable than the other.

5 The scatter graph shows the height and trunk diameter of eight trees.

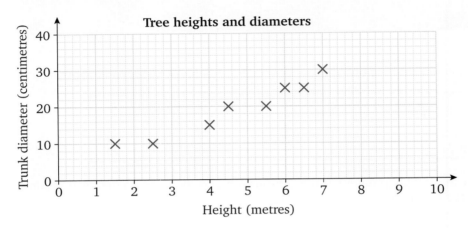

a What is the height of the tallest tree?

b Draw a line of best fit through the points on the scatter graph.

c Describe the relationship shown in the scatter graph.

d **i** Estimate the height of a tree with trunk diameter 35 cm.

    ii Comment on the reliability of your estimate.

6 Adnan is comparing A Level textbooks in order to test the hypothesis:
    'Books with more pages weigh more'.

He records the number of pages and then weighs each textbook.

His results are shown in the table below.

Number of pages	82	90	140	101	160	140	111	152	202
Weight (g)	165	155	210	192	245	96	190	231	280

a One of the readings is an outlier. Which reading is an outlier?
Give a reason why this might occur.

b Is Adnan's hypothesis true or false?
Show your working to justify your answer.

AQA Examination-style questions 🔑

1   The number of hours of sunshine and the maximum temperature at a seaside resort were measured on seven days in June.

Hours of sunshine	5	9	8	6	5	2	4
Temperature (°C)	26	30	29	26	24	19	23

   a   Plot these data as a scatter graph.   *(2 marks)*

   b   Draw a line of best fit on your scatter graph.   *(1 mark)*

   c   Use your line of best fit to estimate the maximum temperature on a day in June when there are 7 hours of sunshine.   *(1 mark)*

   d   Describe the relationship shown by your scatter graph.   *(1 mark)*

   e   Explain why these data may not be representative of the maximum temperatures in June at this seaside resort.   *(1 mark)*

AQA 2007

2   An investigation was carried out by 12 students.
     They counted the number of books in their bags.
     Each student weighed their bag with the books and recorded the total weight.
     The results are shown on the scatter diagram.

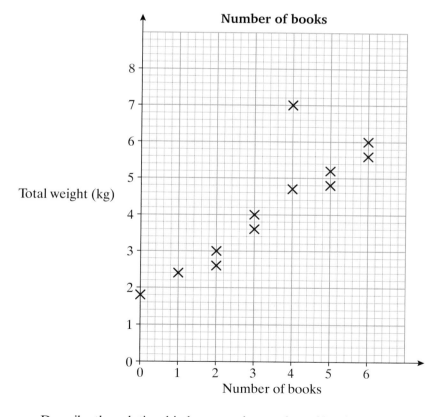

   a   Describe the relationship between the number of books and the total weight.   *(1 mark)*

   b   **i**   Which point does not fit the general pattern?   *(1 mark)*

         ii   If this point is removed from the scatter diagram what effect would this have on the correlation?

              Choose between: Weaker, No effect, Stronger

              Explain your answer.   *(2 marks)*

AQA 2009

9 Probability

Objectives

Examiners would normally expect students who get these grades to be able to:

D

use a two-way table to find a probability

understand mutually exclusive events

identify different mutually exclusive events and know, if they cover all the possibilities, then the sum of their probabilities is 1

C

use probability to estimate outcomes for a population

understand and use relative frequency

B

draw tree diagrams

A

understand independent and non-independent events

find probabilities of successive independent events

A*

find probabilities of successive dependent events.

Key terms

mutually exclusive event
random
theoretical probability
experimental probability
trial
relative frequency

fair
biased
independent events
tree diagram
dependent events
conditional probability
two-way table

Did you know?

Lightning

In the UK around five people are killed by lightning each year. This means that the probability that you will be killed by lightning is approximately $\frac{1}{12\,000\,000}$

In fact, you are more likely to die falling off a ladder (probability $\frac{1}{2\,300\,000}$) or falling out of bed (probability $\frac{1}{2\,000\,000}$) than being killed by lightning.

You should already know:

✔ how to cancel a fraction to its simplest form using a calculator

✔ how to add, subtract and multiply fractions and decimals

✔ how to convert between fractions, decimals and percentages

✔ how to calculate simple probabilities

✔ how to find outcomes systematically

✔ how to use a two-way table to find a probability.

Learn... 9.1 Mutually exclusive events

Mutually exclusive events are events that cannot happen at the same time.

The sum of all the probabilities of mutually exclusive events = 1.

For example, the following events are mutually exclusive.

- Getting a head and getting a tail when a coin is flipped once.
- Getting a three and getting an even number when a dice is rolled once.
- Sleeping and running the marathon at the same time.
- Flying a plane and swimming the Channel at the same time.

For mutually exclusive events, A and B:

P(A or B) = P(A) + P(B). This is known as the OR rule.

P(A) is just a quick way of saying the probability of A, and so on.

So, for example P(Head or Tail) = P(Head) + P(Tail)

Other events are not mutually exclusive and can happen at the same time.

For example, the following events are not mutually exclusive.

- Getting a four and getting an even number when a dice is rolled.
- Getting a red card and getting an ace when a card is taken from a pack (e.g. you could get the ace of hearts).
- Driving a car and listening to the radio.
- Eating a meal and watching TV.

AQA Examiner's tip

In the examination the words 'mutually exclusive' will not be used.

However, you need to understand how events happening at the same time affect probabilities.

Example: 10 discs are placed in a bag.

They are labelled X1, X2, X3, X4, Y1, Y2, Y3, Y4, Z1 and Z2.

One disc is picked at **random**.

Work out the probability of picking the following.

  a   A disc with an X on it

  b   A disc without an X on it

  c   A disc with a 3 on it

  d   A disc without a 3 on it

  e   A disc with an X or a 3 on it

Solution: **a**   Probability of an event happening = $\dfrac{\text{number of outcomes for that event}}{\text{total number of possible outcomes}}$

Probability of picking an X = $\dfrac{4}{10}$ ←— there are 4 discs with an X on
←— there are 10 discs altogether

  b   Probability of an event happening = $\dfrac{\text{number of outcomes for that event}}{\text{total number of possible outcomes}}$

Probability of not picking an X = $\dfrac{6}{10}$ ←— there are 6 discs without an X on
←— there are 10 discs altogether

Here every disc has to be either an X or **not** an X.

The sum of all mutually exclusive probabilities = 1.

The probability of picking an X + the probability of not picking an X = 1.

So, the probability of not picking an X = 1 − the probability of picking an X.

Probability of not picking an X = $1 - \dfrac{4}{10} = \dfrac{6}{10}$

c Probability of an event happening = $\dfrac{\text{number of outcomes for that event}}{\text{total number of possible outcomes}}$

Probability of picking a 3 = $\dfrac{2}{10}$ ← there are 2 discs with a 3 on
← there are 10 discs altogether

d $1 - \dfrac{2}{10} = \dfrac{8}{10}$ (probability of not picking a 3 is 1 − probability of picking a 3)

e The probability of picking a disc with an X or a 3 on is **not** $\dfrac{4}{10} + \dfrac{2}{10}$

This would count the X3 disc twice!

This shows that only mutually exclusive probabilities can be added.

Here you must use the list; there are 5 discs with either an X or a 3 on them. i.e. X1, X2, X3, X4, Y3.

So the probability is $\dfrac{5}{10}$

AQA Examiner's tip

None of the answers in the example are simplified, as simplest form was not asked for.
If the exam question asks you to give an answer in its simplest form, then you must simplify.

Practise... 9.1 Mutually exclusive events

D

1 Which of these pairs of dice events could not happen at the same time?

a Roll a 1 and roll a number less than 5

b Roll a 2 and roll an odd number

c Roll an even number and roll an odd number

d Roll a number more than 3 and a number less than 4

2 The probability that Georgina will wear black on a Sunday is 0.95

What is the probability that Georgina will not wear black on a Sunday?

3 The probability that Mike will have fish and chips for dinner is $\dfrac{7}{100}$

What is the probability that Mike will not have fish and chips for dinner?

4 The probability that Toni will not drink tea at work is 0.001

What is the probability that Toni will drink tea at work?

5 Losalot Town are playing in a football tournament.

Here are some probabilities for the outcome of their opening match.

Complete the table.

Probability of winning	Probability of drawing	Probability of losing
$\dfrac{1}{10}$	$\dfrac{1}{5}$	

C

6 A bag contains coloured discs.

Each disc also has a letter on it.

There are 5 red discs D, E, F, G and H.

There are 8 blue discs D, E, F, G, H, I, J and K.

There are 2 yellow discs D and E.

C

Work out the probability of picking a disc that:

a is red

d is red or has an E on it

b has an E on it

e is yellow or has an F on it

c does not have an E on it

f is blue or has an H on it.

7 The table shows the probabilities that a student will choose a certain drink.

Lemonade	Cola	Orange	Other
0.3	0.2		0.1

What is the probability that the student will choose:

a orange

b cola or lemonade?

! 8 Of the people attending a festival, one is chosen at random to win a prize.

The probability the chosen person is male is 0.515

The probability the chosen person is married is 0.048

The probability the chosen person is a married male is 0.029

What is the probability the chosen person is an unmarried female?

! 9 In a game you choose to throw either one or two ordinary dice.

Your score is the number (if one dice) or sum of the numbers (if two dice).

You need to score a 4 to win the game.

Should you choose to roll 1 dice or 2 dice? Justify your choice.

? 10 A bag contains shapes which are coloured.

The probability of a red square is 0.2

The probability of a red shape is 0.2

Write down what you know about the red shapes in the bag.

Learn... 9.2 Relative frequency

The probabilities so far have all been theoretical probabilities.

Theoretical probability is the probability of an event based on expectation (or theory).

Experimental probability is the probability of an event based on testing (or experiment).

A probability experiment is a test in which a number of **trials** are performed.

The experimental probability is also called the **relative frequency.**

$$\text{Relative frequency of an event} = \frac{\text{number of times an event has happened}}{\text{total number of trials}}$$

Example: Niles rolls an ordinary dice 600 times.

His results are shown in the table.

Score	1	2	3	4	5	6
Frequency	92	107	103	99	97	102

a Work out the relative frequency for each score.

b How many of each score would you expect if the dice was **fair**?

c Do you think the dice is **biased**? Explain your answer.

Solution: **a** Relative frequency of an event $= \dfrac{\text{number of times an event has happened}}{\text{total number of trials}}$

In each case, the total number of trials is 600.

The relative frequencies are shown in the table.

Score	1	2	3	4	5	6
Frequency	92	107	103	99	97	102
Relative frequency	$\dfrac{92}{600}$	$\dfrac{107}{600}$	$\dfrac{103}{600}$	$\dfrac{99}{600}$	$\dfrac{97}{600}$	$\dfrac{102}{600}$

b You would expect **about** 100 of each number if the dice is fair ($\dfrac{1}{6} \times 600$).

c The dice does not look as though it is biased.

The relative frequencies are reasonably close to the theoretical probabilities.

Score	1	2	3	4	5	6
Frequency	92	107	103	99	97	102
Relative frequency	$\dfrac{92}{600}$	$\dfrac{107}{600}$	$\dfrac{103}{600}$	$\dfrac{99}{600}$	$\dfrac{97}{600}$	$\dfrac{102}{600}$
Theoretical probability	$\dfrac{100}{600}$	$\dfrac{100}{600}$	$\dfrac{100}{600}$	$\dfrac{100}{600}$	$\dfrac{100}{600}$	$\dfrac{100}{600}$

Throwing the dice 600 times is a large number so getting values close to 100 ($600 \div 6$) is a sign of no bias.

AQA **Examiner's tip**

Relative frequencies should always be given as fractions or decimals. Giving the frequencies will often score zero.

Example: When Niles rolled the dice, he kept a record of the number of 4s in every 10 throws.

Here are his results.

Number of throws	10	20	30	40	50	60	70	80	90	100	110	120
Number of 4s	1	5	8	10	10	12	13	13	15	15	16	18

a Find the relative frequency after every 10 throws.

b Draw a line graph to show these results.

c What does the graph show about the relative frequency values?

Solution: **a** After 10 throws there had been one 4 giving a relative frequency of $\dfrac{1}{10} = 0.1$

After 20 throws there had been five 4s giving a relative frequency of $\dfrac{5}{20} = 0.25$

AQA **Examiner's tip**

Use decimals to make graph plotting and comparison easier.

Number of throws	10	20	30	40	50	60	70	80	90	100	110	120
Number of 4s	1	5	8	10	10	12	13	13	15	15	16	18
Relative frequency	$\dfrac{1}{10}$	$\dfrac{5}{20}$	$\dfrac{8}{30}$	$\dfrac{10}{40}$	$\dfrac{10}{50}$	$\dfrac{12}{60}$	$\dfrac{13}{70}$	$\dfrac{13}{80}$	$\dfrac{15}{90}$	$\dfrac{15}{100}$	$\dfrac{16}{110}$	$\dfrac{18}{120}$
Relative frequency	0.1	0.25	0.27	0.25	0.2	0.2	0.19	0.16	0.17	0.15	0.15	0.15

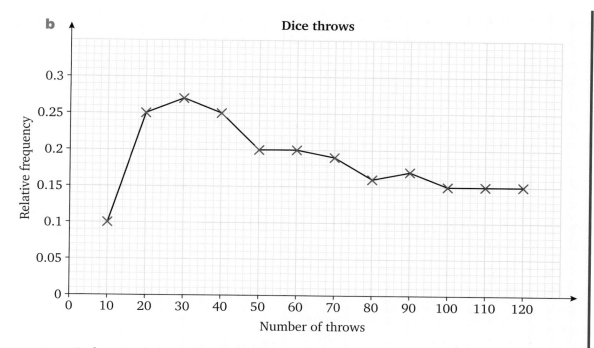

b

Dice throws

c   Early on in the experiment the line is unpredictable.
This shows that using only a few results would be unreliable.

After a while the plotted values are all very similar.
This shows that more results leads to better estimates of probability.

Practise...   9.2 Relative frequency   D C B A A*

1   Ruth flips a coin 240 times.

a   How many times would she expect to get a tail?

b   She actually gets 109 tails.
Do you think the coin is biased?
Give a reason for your answer.

2   Over a long time it is found that the probability of a faulty
light bulb is 0.01

a   How many light bulbs would you expect to be faulty in a
batch of 800?

b   One day, a light bulb checker finds 17 faulty bulbs.
Estimate how many bulbs she has checked that day.

3   The table shows the frequency distribution after drawing a card from a pack
40 times. The card is put back after each draw.

	Results from 40 draws			
	Club	Heart	Diamond	Spade
Frequency	9	9	12	10

a   What is the relative frequency of getting a heart?

b   What is the relative frequency of getting a red card?

c   What is the theoretical probability of getting a club?

d   Ciaron says 'If you drew a card out 80 times you would probably get twice
as many of each suit.' Explain why Ciaron is wrong.

D

C

C

4 Pete has a spinner with coloured sections of equal size.

He wants to know the probability that his spinner lands on blue.

He spins it 100 times and calculates the relative frequency of blue after every 10 spins.

His results are shown on the graph.

a Use the graph to calculate the number of times the spinner landed on blue:

   i after the first 20 spins

   ii after the first 70 spins.

b From the graph, estimate the probability of the spinner landing on blue.

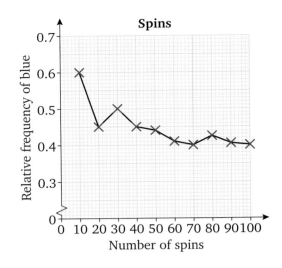

5 Izzy is rolling a dice.

After every 10 rolls she works out the relative frequency of a score of 1.

The diagram shows the relative frequency throughout the experiment.

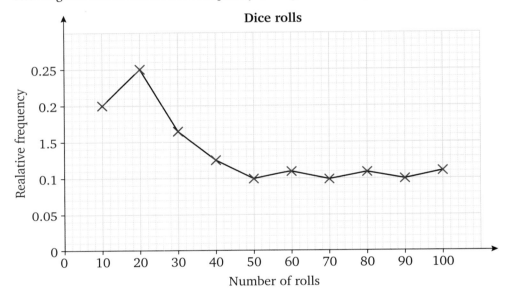

a After 20 rolls, how many times did she roll a 1?

b How many times did she roll a 1 between the 21st and 50th rolls?

c Do you think this dice is biased? Explain your answer.

⚠ 6 The table shows the frequency distribution after rolling a dice 270 times.

Results from 270 rolls of dice					
	1	2	3	4	5	6
Frequency	52	56	41	37	45	39

a What is the relative frequency of getting a 2?

b What is the relative frequency of getting a score greater than 4?

c What is the relative frequency of getting an even number?

d Ellie says the relative frequency of getting a score less than 3 is $\frac{149}{270}$

Is this correct? Explain your answer.

e Which one of the frequencies is the same as the result you would expect from theoretical probability?

7 Archie has a bag of counters.

Inside there are red, green, blue and yellow counters.

He thinks there is the same number of each colour in the bag.

He collects data by picking counters one at a time from the bag.
He replaces each counter in the bag before picking again.

Here are his results.

After 50 picks

Colour	Red	Green	Blue	Yellow
Frequency	20	8	6	16

After 100 picks

Colour	Red	Green	Blue	Yellow
Frequency	30	22	20	28

After 200 picks

Colour	Red	Green	Blue	Yellow
Frequency	57	55	43	45

After 400 picks

Colour	Red	Green	Blue	Yellow
Frequency	104	99	105	92

a Draw the relative frequencies for each colour all on the same chart.

b What evidence is there to support Archie's opinion after:

   i 50 picks   **ii** 100 picks   **iii** 200 picks   **iv** 400 picks

Learn... 9.3 Independent events and tree diagrams

Events are **independent** if the outcome of one event does not affect the outcome of the other.

If two events are independent, then the probability that they will both happen is found by multiplying their probabilities together.

This can be written as P(A and B) = P(A) × P(B). This is known as the AND rule.

So, a dice showing an even number and a coin showing a head **are** independent events. Getting a six on successive throws of a fair dice are also independent events.

A **tree diagram** is a useful tool for showing probabilities.

The probabilities are written on the branches of the tree.

Example: A red dice and blue dice are rolled at the same time.
Find the probability that both dice show a six.

Solution: The scores on the two dice are independent.

The probability of a six on both dice

$$= \text{the probability of a six on a red dice AND a six on the blue dice.}$$

$$= \frac{1}{6} \times \frac{1}{6}$$

$$= \frac{1}{36}$$

Example: 🔢 The probability that Jeff is late for work on any particular day is 0.05

    a   Draw a tree diagram to show the possible outcomes for the two days.

    b   Use the tree diagram to find the probability that on two consecutive days:

        i   Jeff is late on both days

        ii   Jeff is late just once.

Solution: **a**   L = late   N = not late

Day 1	**Day 2**		**Outcome**	**Probability**
		L	LL	$0.05 \times 0.05 = 0.0025$
0.05	0.05 L			
	0.95	N	LN	$0.05 \times 0.95 = 0.0475$
		L	NL	$0.95 \times 0.05 = 0.0475$
0.95	0.05 N			
	0.95	N	NN	$0.95 \times 0.95 = 0.9025$

> AQA *Examiner's tip*
>
> The probabilities on any given pair of branches must add up to 1.
> The total probability of all the final outcomes should also be 1.

    b   **i**   To be late on both days means being late on day 1 **AND** being late on day 2.

            The events are independent so these probabilities can be multiplied.

            Probability of being late on both days = $0.05 \times 0.05 = 0.0025$

        ii   Probability of being late once = late on day 1 **AND** not late on day 2
                                          OR not late on day 1 **AND** late on day 2

                                  = $(0.05 \times 0.95) + (0.95 \times 0.05)$

                                    = $0.0475 + 0.0475$

                                    = 0.095

Remember that:

AND can be associated with multiplying probabilities of independent events.

OR can be associated with adding probabilities of mutually exclusive events.

> *Bump up your grade*
>
> To get a Grade C, you need to know when to use the 'AND' rule and when to use the 'OR' rule.

Practise...

9.3 Independent events and tree diagrams

D C B A A*

B

1   State whether these events are independent.
Give a reason for your answers.

   a   You roll a fair six-sided dice.

   Event A: scoring a 3 on the dice

   Event B: scoring an odd number on the dice

   b   You pick a card from a normal pack and roll a dice.

   Event A: you get a heart

   Event B: you roll a 4

2   Complete the tree diagram.

   Use the tree diagram to find the probabilities of all possible outcomes.

   A bag has 10 balls, 3 with a letter A and 7 with a letter B.

   One ball is taken out at random, replaced and then a second is taken out at random.

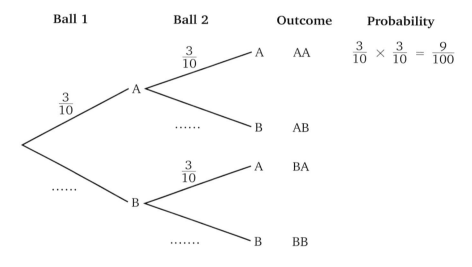

3   The probability that Mike takes a bus to work is 0.8

   Days are independent of each other.

   Work out the probability that Mike takes the bus on Monday and Tuesday.

A

4   Lorraine has a packet of crisps and a fruit juice for lunch every day.

   The probability that she has roast chicken crisps is 0.2

   The probability that she has orange juice is 0.4

   a   What is the probability that she has roast chicken crisps and orange juice?

   b   What is the probability that she does **not** have roast chicken crisps and does **not** have orange juice?

5   The probability that Katie has an argument with her husband on a given day is 0.4

   Days are independent of each other.

   a   Draw a tree diagram to show all the probabilities and outcomes for two days.

   b   Use the tree diagram to find the probability that Katie and her husband:

   i   argue on both days   **iii**   argue on exactly one day

   ii   argue on neither day   **iv**   argue on at least one day.

6 The probability that Jody stays up late on a night before school is 0.15

The probability that Jody stays up late on a night not before school is 0.37

What is the probability that Jodie does not stay up late for 7 consecutive days during a school term?

7 A fair coin is flipped 3 times.

a Draw a tree diagram showing the possible outcomes.

b Use your tree diagram to find the probability that:

i at least two of the coins land on heads

ii no more than one tail occurs.

8 32% of British people can speak a foreign language.

a Two British people are chosen at random.
Use a tree diagram to find the probability that:

i both can speak a foreign language

ii exactly one can speak a foreign language.

b Janet and Mavis are British twins.
Why is it unlikely that the probability they both speak a foreign language is the answer to part **a i**?

c 98% of people born in Luxembourg can speak a foreign language.
What is the probability that 5 randomly chosen people born in Luxembourg can all speak a foreign language?

9 The probability that Kels works 7 days in a week is x.

Weeks are independent of each other.

The probability that Kels works 7 days for two consecutive weeks is 0.1764

Find x.

10 Explain why the probability of two independent events happening is always less than or equal to the individual probabilities.

Learn... 9.4 Dependent events and conditional probability

Events are **dependent** if the outcome of one event affects the outcome of the other. This is also known as **conditional probability**.

Many examples of conditional probability involve choosing items and not replacing them, e.g. counters from bags, students from classes, etc.

Again, a tree diagram is a useful tool for showing probabilities.

As before, the probabilities are written on the branches of the tree.

> **AQA** *Examiner's tip*
>
> The examination uses the term 'without replacement' to indicate that the first item is not put back before the second item is chosen.

Example: A bag contains 10 red discs and 8 blue discs.

One disc is chosen at random and **not** replaced.

A second disc is then chosen at random.

a Work out the probability that both discs are blue.

b Use a tree diagram to show all the possible events and their probabilities.

Solution: **a** The probability that both discs are blue

= probability 1st disc is blue AND 2nd disc is blue

= probability 1st disc is blue × probability 2nd disc is blue

$= \dfrac{8}{18} \times \dfrac{7}{17}$ ← only 7 blue discs left as the 1st one is not replaced

   ← only 17 discs left altogether as the 1st one is not replaced

$= \dfrac{56}{306} = \dfrac{28}{153}$

> **AQA** *Examiner's tip*
>
> If the exam question asks you to give an answer in its simplest form, then you must cancel. Always show the original fraction in case you make a mistake cancelling down.

b

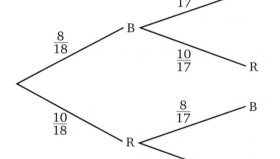

Disc 1	Disc 2	Outcome	Probability
	$\frac{7}{17}$ B	BB	$\frac{8}{18} \times \frac{7}{17} = \frac{56}{306}$
$\frac{8}{18}$ B			
	$\frac{10}{17}$ R	BR	$\frac{8}{18} \times \frac{10}{17} = \frac{80}{306}$
	$\frac{8}{17}$ B	RB	$\frac{10}{18} \times \frac{8}{17} = \frac{80}{306}$
$\frac{10}{18}$ R			
	$\frac{9}{17}$ R	RR	$\frac{10}{18} \times \frac{9}{17} = \frac{90}{306}$

> **AQA** *Examiner's tip*
>
> The probabilities on any given pair of branches must add up to 1.
> The total probability of all the final outcomes should also be 1.

9.4 Dependent events and conditional probability

Practise...

1   A set of cards is numbered 1, 2, 3, 4, 5, 6, 7, 8, 9 and 10.

Given that the first card picked is even, work out the probability of:

a a 7 **b** an 8 **c** a 7 or an 8 **d** not an 8 **e** not a 7

2   A bag contains 10 counters; 6 are black, 4 are white.
Two counters are chosen at random without replacement.

Draw a tree diagram and use it to work out the probability that:

a both counters are black   **c** the counters are of different colours.

b both counters are white

3   A box contains 3 red pencils and 8 blue pencils.
Two pencils are taken from the box without replacement.

Draw a tree diagram and use it to find:

a the probability that both pencils are red

b the probability that both pencils are green

c the probability that one pencil is red and one is green.

A*

4   The cards from Question 1 are face down on the table and in a random order.

One card is picked at random.
It is then put back and the cards are shuffled.
A second card is then picked at random.

a   Work out the probability that one card is a 7 and the other is even.

Starting again from the same set of cards, two cards are now picked at random without replacement.

b   What is the probability that one card is a 7 and the other is even?

5   The probability that Mr Metcalf sleeps through his alarm is 0.09

If he sleeps through his alarm the probability he will miss breakfast is 0.72

If he does not sleep through his alarm the probability he will miss breakfast is 0.26

a   Draw a tree diagram to show these probabilities.

b   Use the tree diagram to find the probability that Mr Metcalf:
  i   sleeps through his alarm and misses breakfast
  ii   does not sleep through his alarm and does not miss breakfast.

! 6   A bag contains 90 balls coloured red or green in the ratio 4 : 5

Two balls are drawn at random.

Using a tree diagram, find the probability that one ball of each colour is picked.

⚙ 7   The National Lottery is played by choosing 6 numbers from the values 1–49 inclusive.

What is the probability of choosing all 6 numbers correctly?

Give your answer correct to 3 s.f.

? 8   The histogram shows the time taken for customer calls to be answered at a call centre.

Two of these customers are chosen at random to answer a survey.

a   Work out the probability that:

  i   both customers waited for more than 4 seconds to be answered

  ii   one customer waited less than 4 seconds and one customer waited for more than 4 seconds.

b   Given that both customers actually had to wait under 3 seconds, what is the probability they both had to wait under 2 seconds?

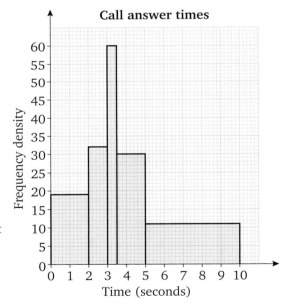

Call answer times

Frequency density vs Time (seconds)

9   The **two-way table** shows information about the gender of performers in a school show.

Two of the performers are chosen at random to talk to the local newspaper.

What is the probability that they are both:

a   boys   **b**   dancers?

	Boys	Girls
Singers	4	12
Dancers	10	16

? 10   Jack has some socks in his drawer. Some of them are black.

If he chooses two socks at random, the probability he gets a black pair is $\frac{132}{380}$

How many socks were in the drawer and how many of them were black?

9 Assess

1 In the UK the probability of being left handed is about 0.11

How many left-handed people would you expect to find in the following?

a a class of 33 children.

b a street of 132 people.

c a town of 55 000 people.

2 Two pentagonal spinners, each with the numbers 1 to 5, are spun.
Their outcomes are added together to give a score.

a Draw a two-way table for the two spinners.

b Use your diagram to find:

    i the probability of a score of 4   **iii** the probability of a score of 9

    ii the probability of a score of 5   **iv** the most likely score.

c Repeat parts **a** and **b** for a score that is the outcomes **multiplied** together.

3 A fair six-sided dice is thrown 250 times and the following results obtained.

Score	1	2	3	4	5	6
Frequency	45	48	43	40	38	36

a What is the relative frequency of a score of 1?

b What is the relative frequency of a score of 6?

c What is the relative frequency of scoring more than 3?

d How does this data confirm that the dice is fair?

e Draw a new table with possible frequencies if this dice was thrown 6000 times.

4 The table below shows the probabilities of selecting tickets from a bag.

The tickets are coloured yellow, black or green and numbered 1, 2, 3 or 4.

	1	2	3	4
Yellow	$\frac{1}{20}$	$\frac{1}{16}$	$\frac{3}{40}$	$\frac{1}{8}$
Black	$\frac{1}{10}$	$\frac{3}{40}$	0	$\frac{3}{40}$
Green	0	$\frac{1}{8}$	$\frac{3}{16}$	$\frac{1}{8}$

A ticket is taken at random from the bag.

Calculate the probability that:

a it is black and numbered 4   **c** it is not black

b it is green   **d** it is yellow or numbered 3

5 The probability Callum has to work overtime on a given day is 0.12

a What is the probability that he has to work overtime on two consecutive days?

b What assumption did you make to answer part **a**?

A

6   The relative frequency of rain for the last 100 hundred years in London is 0.42 for any given day.

   a   Use a tree diagram to show the outcomes for two consecutive days and their probabilities.

   b   Using this data, what is the probability that it will rain on two consecutive days?

   c   Why is this probability unlikely to be correct?

A*

7   Maria plays tennis better on grass courts than any other.

On grass courts she has a 75% chance of winning.

On other courts she has a 60% chance of winning.

She plays 45% of her games on grass courts.

Use a tree diagram to find the probability that she will win the next game she plays.

8   At a party there are 40 balloons for the children to take home.

The table shows the number of each colour.

The balloons are given out at random.

Linda and Natalie are the first two to get a balloon.

Find the probability that neither Linda nor Natalie get a red balloon.

Colour	Number
Blue	12
Green	6
Orange	10
Red	7
Yellow	5

9   There are 12 sandwiches left in a cabinet at a café.

Six are meat sandwiches, four are fish, and the other two are vegetarian.

Earl chooses two sandwiches. His choices are independent. He likes all the choices.

   a   Draw a tree diagram to show the possible outcomes and their probabilities.

   b   What is the probability that exactly one of his sandwiches is a fish sandwich?

AQA Examination-style questions

1   At the end of a training course candidates must take a test in order to pass the course.
The probability of passing the test at the first attempt is 0.8
Those who fail re-sit once.
The probability of passing the re-sit is 0.5
No further attempts are allowed.

   a   **i**   Complete the tree diagram, which shows all the possible outcomes.

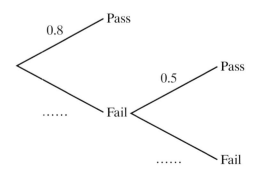

      ii   What is the probability that a candidate fails both attempts and so fails the course?   *(2 marks)*

   b   What is the probability that a candidate passes the course?   *(1 mark)*

   c   Hassan and Louise both take the training course.
What is the probability that one of them passes and one of them fails?   *(3 marks)*

AQA 2008

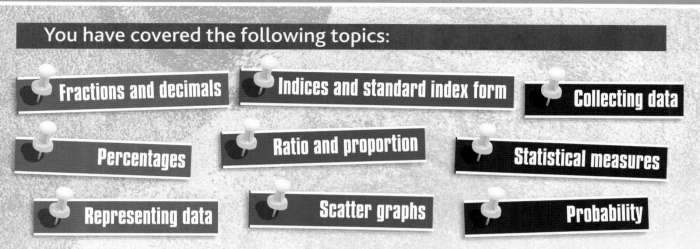

You have covered the following topics:

- Fractions and decimals
- Indices and standard index form
- Collecting data
- Percentages
- Ratio and proportion
- Statistical measures
- Representing data
- Scatter graphs
- Probability

All these topics will be tested in this chapter and you will find a mixture of problem solving and functional questions. You won't always be told which bit of maths to use or what type a question is, so you will have to decide on the best method, just like in your exam.

Example: The cumulative frequency diagrams show the times taken by 50 boys and 50 girls to run 100 metres.

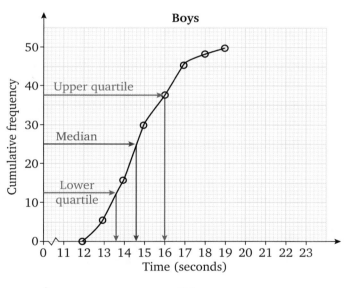

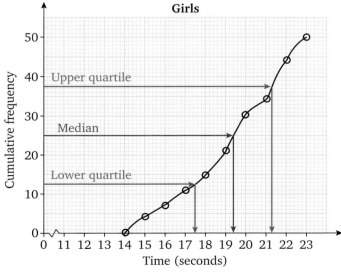

Compare the times taken by the boys and girls. *(6 marks)*

Solution:

The median

The median value is usually the $\frac{1}{2}(n + 1)$th value.

In this case, $n = 50$ for both boys and girls.

Because n is large, the median value can be estimated from the $\frac{1}{2}n$th value (in this case the 25th value).

The median is approximately the time taken by the 25th runner.

The median time for the boys = **14.6 seconds**

The median time for the girls = **19.4 seconds**

Lower quartile

Again, because n is large, the lower quartile (LQ) can be estimated from the $\frac{1}{4}n$th value (the 12.5th value).

The lower quartile is half way between the times taken by the 12th and 13th runner.

Boys LQ = **13.6 seconds**

Girls LQ = **17.5 seconds**

Upper quartile

In a similar way, the upper quartile (UQ) can be estimated from the $\frac{3}{4}n$th value (the 37.5th value).

So the upper quartile is halfway between the times taken by the 37th and 38th runner.

Boys UQ = **16 seconds**

Girls UQ = **21.3 seconds**

Inter-quartile range

The inter-quartile range (IQR) is the difference between the upper and lower quartiles.

IQR = UQ − LQ

Boys IQR = **16 − 13.6 = 2.4 seconds**

Girls IQR = **21.3 − 17.5 = 3.8 seconds**

Summary of measures

Boys: Median = **14.6 seconds**, IQR = **2.4 seconds**

Girls: Median = **19.4 seconds**, IQR = **3.8 seconds**

Comparing the data sets

The average time for the boys is smaller than for the girls. This means the boys are **on average** faster than the girls.

The inter-quartile range for the girls was greater than that for the boys. This means that the girls run **more variable** times than the boys.

Mark scheme
- 2 marks for calculating the median for the boys and the girls.
- 2 marks for calculating the inter-quartile range for the boys and the girls.
- 1 mark for saying that the boys on average are faster than the girls.
- 1 mark for saying that the girls run more variable times than the boys.

AQA Examiner's tip

When you are asked to compare data sets make two statements. One comparing average values and one comparing measures of spread.

Average values

Use the word average in your statement. For example, 'boys run faster than girls' is wrong because boys only run faster on average.

Measures of spread

Show in your statement that you know that a smaller measure of spread means that the values in the data set are closer together. Useful phrases for this are 'less variable' and 'more consistent'.

Example: A council tax bill will be reduced by 5% if full payment is made before the end of April.

a Mr Akram paid £1350 on 1 May.

How much would he have paid if he had paid the previous day? *(3 marks)*

b Miss Wise paid £1539 on 27 April.

How much would she have had to pay four days later? *(3 marks)*

Solution: **a** To reduce by 5%, multiply by 0.95

1350×0.95

$= £1282.50$

Mark scheme
- 1 mark for using 95%
- 1 mark for multiplying by 0.95.
- 1 mark for the final correct answer.

AQA Examiner's tip

Use multipliers whenever possible. This is not the only way to reduce by 5% but when you are allowed a calculator it is the easiest method.

The calculator gave the answer as 1282.5. Remember that money should have two digits for pence. £1282.5 would not get full marks.

b £1539 is the amount that includes the 5% reduction.

So, $95\% = £1539$

$1\% = £1539 \div 95$

$= £16.2$

$100\% = £16.2 \times 100$

$= £1620$

Mark scheme
- 1 mark for using 95%
- 1 mark for working out 1% then multiplying by 100.
- 1 mark for the final correct answer.

AQA Examiner's tip

This is known as a reverse percentage problem.
A very common error is to increase £1539 by 5%

Alternative Solution: **b** You can use a multiplier to find the reverse percentage in one easy step.

Divide by the multiplier:

$£1539 \div 0.95 = £1620$

Mark scheme
- 1 mark for using 0.95
- 1 mark for dividing by 0.95
- 1 mark for the final correct answer.

AQA Examiner's tip

For both parts you should check your answers.

In part **a** Mr Akram would have paid less if he had paid earlier so check that your answer is less than £1350.

In part **b** Miss Wise would have paid more if she had paid later so check that your answer is more than £1539.

D

1 Twenty miles per hour speed limit signs are put up on a housing estate.
The stem-and-leaf diagrams show the speeds of 15 cars both before and after
putting up the signs.

Before

1	9	9			
2	4	6	9	9	
3	0	0	2	6	6
4	0	1	2		
5	2				

After

1	6	8	8	9	9	9
2	0	2	3	4	6	
3	2	5				
4	2					
5	1					

Key: 3 | 5 means 35 mph

Compare the speeds of the cars before and after putting up the signs.

2 School caterers ask pupils these questions.

a What is your age?

b How often do you buy a vegetarian meal?

c Pasta is healthier than chips.
Do you agree that we should stop
serving chips?

Write a criticism of each of the questions asked.

3 A mobile phone in Costlo is priced at £130.

The same mobile phone in Tescbury is priced at £155.

COSTLO PRICES EXCLUDE VAT

TESCBURY PRICES INCLUDE VAT

VAT rate is 17.5%

Compare the costs of the phone in the two shops.

4 The scatter graph shows the marks for ten students in French and German exams.

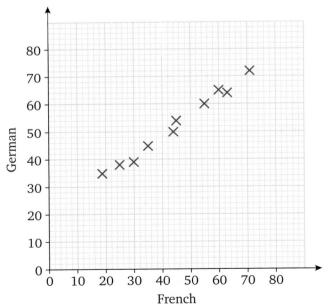

a Students who scored a total mark of 120 or more in both exams combined were
awarded a certificate.
How many of the students were awarded a certificate?

b Sally was ill and missed the French exam but scored 60 marks in the
German exam. Should Sally get a certificate? Explain your answer.

5 Here is a recipe for cakes.

Sara wants to make some cakes for a children's party.

She decides to make 18 small cakes for the under-5s and 30 larger cakes for the older children.

Write a list showing the amounts of each ingredient that she needs.

You must show your working.

> **Ingredients for 24 small cakes or 12 larger cakes**
>
> 125 g of castor sugar
>
> 125 g of softened butter
>
> 125 g of self-raising flour
>
> 2 large eggs, lightly beaten
>
> 1 teaspoon of vanilla extract
>
> 2 tablespoons of milk

6 **a** Jenny spends $\frac{3}{8}$ of the money in her purse.
  She has £17.65 left in her purse.
  Work out the amount of money that Jenny spends.

  b A team have won $\frac{2}{5}$ of their matches and drawn $\frac{1}{4}$ of them.
  The team has lost 7 games.
  A team scores 3 points for a win and 1 point for a draw.
  A team scores no points when a game is lost.
  How many points has the team scored?

7 In an experiment this trial was carried out a number of times.

> Take a bead at random from a bag containing 200 beads.
>
> Record the colour of the bead and then put it back in the bag.

The table shows the results from this experiment after 50 trials and after 300 trials.

Number of trials	Number of red beads obtained
50	11
300	78

  a Calculate the relative frequency of the number of red beads after 50 trials.

  b Work out the most likely number of red beads in the bag.
  Show working to justify your answer.

8 The Angel Falls in Venezuela is 780 metres tall (to the nearest 10 metres).

What is its smallest possible height?

C

9 The number of pupils at a school increases from 750 to 780.

Work out the percentage increase.

10 PQRS is part of a number line.

PQ : QR = 5 : 7

PR : RS = 4 : 1

P = 10

S = 55

Find the value of Q.

11 Each term Zoe takes tests in English, French and German. Last term her mean mark in the tests was 44.

Her target is to reach a mean mark of 50.

This term her mark in English improves by 10, but her mark in French decreases by 4.

By how much must she improve her German mark to reach the target?

12 Adam was saving his money for a games console that he had seen advertised in his local computer shop at a price of £280.

When he had saved exactly £280 he went to the shop to buy the games console.

The shopkeeper told him that the price had increased by 8% the previous week.

However, the shop had that day started a sale with a $7\frac{1}{2}$% reduction on all games consoles.

Did Adam have enough money to buy the games console?

You **must** show your working.

C
B

13 In a game, players throw two ordinary fair dice.
One of the dice is red and the other is blue.

Players can choose one of these options to obtain their score.

Option 1	**Option 2**
Add the numbers on the two dice.	Ignore the number on the red dice and double the number on the blue dice.

a Tom throws a 3 on the red dice and a 4 on the blue dice.
What are his possible scores?

b To win the game Charles needs a score of 10. Show that the probability Charles scores 10 on his next turn is $\frac{2}{9}$

c Charles actually throws 4 on the red dice and 1 on the blue dice.
Should Charles choose Option 1 or Option 2 to give himself the best chance of winning on his following move?

You **must** show working to justify your answer.

B

14 **a** Work out $10^2 + 10^1 + 10^0 + 10^{-1}$

b Work out $9 \times 10^{-2} + 8 \times 10^{-1} + 7 \times 10^0 + 6 \times 10^1 + 5 \times 10^2$

15 **a** Put these numbers in order of size. Start with the largest.

1.5×10^{-1} 2^{-3} 3^0 $0.25^{\frac{1}{2}}$

b The thickness of each page of a book is 5×10^{-3} centimetres.
The thickness of the front cover is 1.2 millimetres.
The thickness of the back cover is 1.2 millimetres.
The book has 70 pages.

Work out the thickness of the book.
Give your answer in centimetres.

16 **a** One person produces about 9×10^9 red blood cells in one hour.
How many million blood cells are produced by one person in one hour?

b The length of a bacteria cell is 2.5×10^{-4} millimetres.
The length of a red blood cell is 30 times the length of a bacteria cell.

Work out the length of a red blood cell.
Give your answer in standard form.

17 The map and table give some information about the continents of the world.

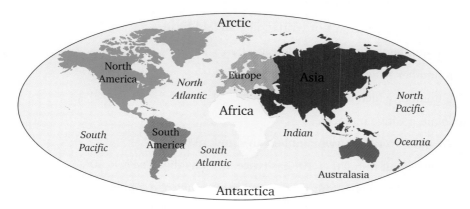

Continent	Population	Area (m²)
Europe	7.27×10^8	9.94×10^9
Asia	3.88×10^9	4.46×10^{10}
North America	5.02×10^8	2.43×10^{10}
South America	3.80×10^8	1.78×10^{10}
Africa	8.78×10^8	3.01×10^{10}
Australasia/Oceania	3.20×10^7	7.69×10^9

$$\text{Population density} = \frac{\text{Population}}{\text{Area}}$$

The table shows **all** of the populated continents.

Work out the average population density of all of the populated continents.

You **must** show your working.

18 There are 30 students in a class. The mean number of pets that they own is 0.8
No one owns more than 4 pets.

Copy and complete the table.

Number of pets	0	1	2	3	4
Number of students		12		2	1

B

19 Della buys a toaster for £35.20 in a sale. In the sale, prices are reduced by 20%

How much did the toaster cost before the sale?

20 A decorating company uses five workers to decorate six rooms in three days.
They have another job decorating eight different rooms of the same size.

They want to complete this job in two days.

How many workers should they use?

B
A

21 Here is a list of numbers.

4×10^7

8.5×10^{-3}

3.6×10^5

0.9×10^4

2×10^6

4.9×10^{-1}

a From this list write down:

   i the number which is **not** written in standard form

   ii the number which is the smallest

   iii a number which is greater than 1 million

   iv a number which is a square number.

b a and b are positive integers.

$a \times 10^b$ and $b \times 10^a$ are two numbers in standard form.

When written out in full $a \times 10^b$ has twice as many digits as $b \times 10^a$.

Show that there are only four possible pairs of values of a and b.

22 The table shows the percentage annual salary increase that a company has given to its employees over the past few years.

Year	2006	2007	2008	2009	2010
% salary increase	3%	6%	2.5%	1%	0%

a Before the 2008 increase John's salary was £25 000.
What is his salary in 2010?

b Carol joined the company in 2006. She did not receive the salary increase that year.
She left the company after the 2008 increase with a salary of £46 176.25.

   i Show that the percentage increase in her salary during the time she was working for the company was 8.65%

   ii Work out the salary she started with in 2006.
You **must** show your working.

A

23 Bruce ran for 21.8 seconds at an average speed of 7 metres per second.

The time is correct to the nearest tenth of a second.

The speed is correct to the nearest metre per second.

Work out the maximum distance that Bruce could have travelled.
Give your answer in metres.

24 Box A contains one blue and two red counters.
Box B contains two blue and one red counter.

One counter is taken at random from each box.

What is more likely:
the two counters are the same colour or the two counters are a different colour?

You **must** show working to justify your answer.

25 One hundred students run in a cross-country race.
The table shows the distributions of their times.

Time, t (minutes)	$20 \leqslant t < 25$	$25 \leqslant t < 30$	$30 \leqslant t < 45$	$45 \leqslant t < 60$	$60 \leqslant t < 70$
Frequency	4	12	42	33	9

a Draw a histogram to represent this information.

b The fastest 30 students receive a medal.
Estimate the time taken by the slowest student who receives a medal.

26 The table shows the number of people working different shifts at a factory.

	Age	Shift		
		am	pm	Evening
Male	Under 30	6	8	9
	30 and over	7	12	17
Female	Under 30	8	11	18
	30 and over	6	15	23

A male worker is chosen at random and a female worker is chosen at random.

Work out the probability that they both work on the same shift.

AQA Examination-style questions 🔵

1 A train company surveys opinions about the quality of its service.
On a particular train there are 140 passengers travelling standard class and 35 passengers travelling first class.
A sample of 40 passengers is taken, stratified according to the class of travel.

a Give one advantage of using a stratified sample in this situation. *(1 mark)*

b Calculate the number of passengers travelling standard class and the number of passengers travelling first class that should be in the sample of 40. *(3 marks)*

c Give another way in which the sample of passengers could be stratified. *(1 mark)*

AQA 2009

2 The number of mince pies sold by a bakery increases by 60% in December compared to November.
The number of mince pies sold in January is the same as the number sold in November.
Work out the percentage decrease in sales for January compared to December. *(3 marks)*

AQA 2008

Glossary

amount – the principal + the interest (i.e. the total you will have in the bank or the total you will owe the bank, at the end of the period of time).

average – the name given to a single value that represents a set of data.

back-to-back stem-and-leaf diagram – a stem-and-leaf diagram where the stem is down the centre and the leaves from two distributions are either side for comparison.

balance – (i) how much money you have in your bank account or (ii) how much you owe a shopkeeper after you have paid a deposit.

biased – in the context of probability, not having the expected chance of happening.

census – an official count or survey.

class interval – the range of values within a group (class) of grouped data.

closed questions – questions that control the responses allowed by using option boxes.

compound interest – the interest paid is added to the amount in the account so that subsequently interest is calculated on the increased amount; the amount of money in the account grows exponentially.

conditional probability – the probability of an event (A), given that another (B) has already occurred.

continuous data – quantitative data that are measured but must be rounded to be recorded such as heights, weights, times.

controlled experiment – data collection by a planned investigation of some type such as checking heart rates of runners.

coordinates – a system used to identify a point; an x-coordinate and a y-coordinate give the horizontal and vertical positions.

correlation – a measure of the relationship between two sets of data; correlation is measured in terms of type and strength.

credit – when you buy goods 'on credit' you do not pay all the cost at once; instead you make a number of payments at regular intervals, often once a month. NB when your bank account is in 'credit', this means you have some money in it.

cumulative frequency – the total frequency up to and including a particular value.

cumulative frequency diagram – the name given to any diagram that shows the cumulative frequencies for a distribution.

data classes – a range of data values grouped together.

data collection sheet – see **observation sheet**.

data logging – automatic collection of data by a 'dumb' machine, e.g. in a shop or car park entrance.

denominator – the bottom number of a fraction, indicating how many fractional parts the unit has been split into; examples: in the fractions $\frac{2}{7}$, $\frac{3}{100}$, $\frac{7}{9}$ the denominators are 7 (indicating that the unit has been split into 7 parts, which are sevenths) 100 and 9.

dependent events – events are dependent if the probability of the outcomes for the second event are changed by the outcome of the first event.

deposit – an amount of money you pay towards the cost of an item; the rest of the cost is paid later.

depreciation – a reduction in value (of used cars, for example).

discount – a reduction in the price. Sometimes this is for paying in cash or paying early.

discrete data – quantitative data taking exact values such as frequencies, shoe size, dice scores.

equivalent fractions – two or more fractions that have the same value; equivalent fractions can be made by multiplying or dividing the numerator and denominator of any fraction by the same number; examples: the fractions $\frac{1}{3}$, $\frac{2}{6}$, $\frac{5}{15}$ are equivalent and all have the same value.

event – something that takes place that we want to find the probability of. For example, for finding the probability of 'getting an even number with one throw of a dice', the event is 'getting an even number with one throw of a dice'.

experimental probability – the chance of a particular outcome based on results of experiments or previous data.

exponent – see **index**.

fair – without bias e.g. a fair coin has an equal chance of falling on heads or tails.

frequency diagram – any chart or diagram which compares the frequencies of objects.

frequency distribution – shows the number of times particular values have occurred.

frequency polygon – a frequency diagram for continuous data with a line joining the midpoints of the class intervals using the appropriate frequencies.

frequency table – a table showing total number (frequency) against data values. Like a tally chart but with a number instead of tallies.

grouped data – data that are separated into data classes.

histogram – a diagram for continuous data with bars as rectangles whose areas represent the frequency.

hypothesis – an idea that is put forward for investigation; for example, 'More girls are left handed than right handed'. Data would be collected and analysed in order to investigate whether the hypothesis might be true or not.

independent events – two events are independent if the outcome of the second is not affected by the outcome of the first.

index – the index (or power or exponent) tells you how many times the base number is to be multiplied by itself.

index (or power or exponent)

5^3

Base

indices – the plural of index: see **index**.

integer – a whole number, positive, negative or zero. Examples: 10, −8, +163.

interest – the money paid to you by a bank or building society when you save your money in an account with them. NB it is also the money you pay for borrowing from a bank.

inter-quartile range – the upper quartile minus the lower quartile.

line graph – a diagram for continuous data, usually over a period of time.

line of best fit – a line drawn to represent the relationship between two sets of data; ideally it should only be drawn where the correlation is strong, for example:

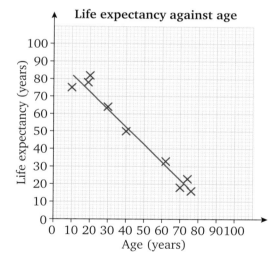

Life expectancy against age

lower bound – the lower limit on the possible size of a measurement and the lowest possible value it can take; for example, if a length is measured as 62 cm to the nearest centimetre the lower bound of the length is 61.5 cm.

lower quartile – the lower quartile (Q_1) is the value $\frac{1}{4}$ along a set of data.

mean – the total of all the values divided by the number of values (also called the arithmetic mean).

$$\text{Mean} = \frac{\text{the total of (frequencies} \times \text{values)}}{\text{the total of frequencies}} = \frac{\Sigma fx}{\Sigma f}$$

median – the middle value when the data are listed in order.

mixed number – a fraction that has both a whole number and a fraction part.

modal class or modal group – the class or group within a frequency table that occurs most often.

mode – the value or item that occurs most often.

mutually exclusive events – events that are mutually exclusive cannot happen at the same time e.g. a 4 and an odd number.

negative correlation – as one set of data increases, the other set of data decreases.

numerator – the top number of a fraction, indicating how many parts there are in the fraction; examples: in the fractions $\frac{4}{5}, \frac{23}{26}, \frac{6}{15}$ the numerators are 4, 23 and 6.

observation – data collection by watching something happen.

observation sheet – prepared tables to record responses to questionnaires or outcomes for an observation such as noting car colours.

open questions – allow for any response to be made by using an answer space.

outcome – one of the possible results of an experiment or trial. For example, when rolling a dice there are six possible outcomes: 1, 2, 3, 4, 5, 6.

outlier – a value that does not fit the general trend, for example:

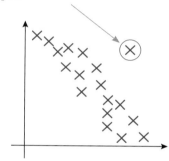

percentage – 'a number of hundredths of' as in 15% means $\frac{15}{100}$

pilot survey – a small scale survey carried out before the main survey.

population – every possible item that could occur in a given situation.

positive correlation – as one set of data increases, the other set of data increases.

power – see **index**.

primary data – data that are collected specifically to answer the research question.

principal – the initial amount of money put into the bank (or borrowed from the bank).

proportion – compares one part with the whole, whereas a ratio compares one part with another. If a class has 10 boys and 15 girls, the proportion of boys in the class is $\frac{10}{25}$ (which simplifies to $\frac{2}{5}$). The proportion of girls in the class is $\frac{15}{25}$ (which simplifies to $\frac{3}{5}$).

proportional change – a change in direct proportion to the amount being changed; if the amount being changed is doubled, the change is twice as big.

qualitative data – data that cannot be measured using numbers e.g. hair colour, sports, breeds of sheep.

quantitative data – data that can be measured such as heights, ages, times, frequencies.

quartiles – the $\frac{1}{4}(n + 1)$th value and the $\frac{3}{4}(n + 1)$th value in a data set of n items.

questionnaire – data collection by a series of questions requiring responses.

random – outcomes are random if they have an equal or set probability but otherwise cannot be predicted.

random sampling – every member of the population has an equal chance of being in the sample.

range – the difference between the highest value and the lowest value in a distribution (a measure of spread, not a measure of average).

rate – the percentage at which interest is added.

ratio – a means of comparing numbers or quantities. A ratio shows how much bigger one number or quantity is than another. If two numbers or quantities are in the ratio 1 : 2, the second is always twice as big as the first. If two numbers or quantities are in the ratio 2 : 5, for every 2 parts of the first there are 5 parts of the second.

raw data – data before they have been sorted in any way.

relative frequency – the fraction or proportion of the number of times out of the total that a particular outcome occurs.

round – give an approximate value of a number; numbers can be rounded to the nearest 1000, nearest 100, nearest 10, nearest integer, significant figures, decimal places, etc.

sample – a small part of a population from which information is taken.

sample size – the number of people or items in the sample.

scatter graph – a graph used to show the relationship between two sets of variables, for example, temperature and ice cream sales.

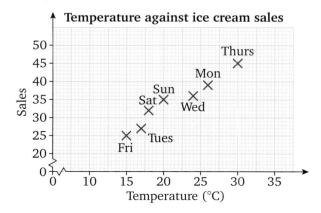

secondary data – data that others have collected; anything from newspapers, the internet and similar sources.

significant figure – a digit in a number that is significant in the accuracy of the number. The closer the digit is to the beginning of the number the greater its significance. Zeros can be significant figures but are often in a number just to maintain the correct place value. Examples: The number 30 597 when rounded to three significant figures is 30 600; the first zero is significant but the final two are not. The number 3.0587 rounded to three significant figures is 3.06

standard index form – standard index form is a shorthand way of writing very large and very small numbers.

stem-and-leaf diagram – a frequency diagram which uses the actual values of the data split into a stem and leaves with a key.

stratified (random) sampling – if the population falls into a series of groups or 'strata' this ensures that the sample is representative of the population as a whole; for example, if the population has twice as many boys as girls, then the sample should have twice as many boys as girls (individuals within each strata are then chosen using random sampling).

survey – general name for data collection using interviews or questionnaires.

tally chart – a method of organising raw data into a table using a five bar gate method of tallying.

theoretical probability – the chance of a particular outcome based on equally likely outcomes.

tree diagram – a diagram used to calculate probabilities of combined events.

trial – a probability experiment consisting of a number of individual trials. For example, if an experiment is to 'throw a dice' and it is thrown 20 times, then that is 20 trials.

two-way table – a table showing information about two sets of data at the same time.

unitary method – a method of calculating quantities that are in proportion by first finding one unit.

unitary ratio – a ratio in the form 1 : n or n : 1; this form of ratio is helpful for comparison, as it shows clearly how much of one quantity there is for one unit of the other.

upper bound – the upper limit on the possible size of a measurement; for example, if a length is measured as 62 cm to the nearest centimetre, the upper bound of the length is 62.5 cm.

upper quartile – the upper quartile (Q_3) is the value $\frac{3}{4}$ along a set of ordered data.

VAT (Value Added Tax) – this tax is added on to the price of goods or services.

zero or no correlation – where there is no obvious relationship between the two sets of data.

Index